Erling B. Andersen

Introduction to the Statistical Analysis of Categorical Data

With 16 Figures
and 121 Tables

 Springer

Prof. Erling B. Andersen
University of Copenhagen
Department of Statistics
6 Studiestræde
DK-1455 Copenhagen
Denmark

ISBN 3-540-62399-X Springer-Verlag Berlin Heidelberg New York

Cataloging-in-Publication Data applied for
Die Deutsche Bibliothek - CIP-Einheitsaufnahme
Andersen, Erling B.: Introduction to the statistical analysis of categorical data analysis: with 121 tables / Erling B. Andersen. - Berlin; Heidelberg; New York; Barcelona; Budapest; Hong Kong; London; Milan; Paris; Santa Clara; Singapore; Tokyo: Springer, 1997
  ISBN 3-540-62399-X

This work is subject to copyright. All rights are reserved, whether the whole or part of the material is concerned, specifically the rights of translation, reprinting, reuse of illustrations, recitation, broadcasting, reproduction on microfilm or in any other way, and storage in data banks. Duplication of this publication or parts thereof is permitted only under the provisions of the German Copyright Law of September 9, 1965, in its current version, and permission for use must always be obtained from Springer-Verlag. Violations are liable for prosecution under the German Copyright Law.

© Springer-Verlag Berlin · Heidelberg 1997
Printed in Germany

The use of general descriptive names, registered names, trademarks, etc. in this publication does not imply, even in the absence of a specific statement, that such names are exempt from the relevant protective laws and regulations and therefore free for general use.

SPIN 10547290   42/2202-5 4 3 2 1 0 - Printed on acid-free paper

Introduction to the Statistical Analysis of Categorical Data

Springer
*Berlin
Heidelberg
New York
Barcelona
Budapest
Hong Kong
London
Milan
Paris
Santa Clara
Singapore
Tokyo*

To my son Søren

Preface

This book is an introduction to categorical data analysis intended as textbook for a one-semester course of approximately 50 teaching units.

The book is a short version of my book The Analysis of Categorical Data Analysis. 3.ed. 1994.

It consists of 8 chapters covering, in addition to an introduction and a chapter on basic theory, six of the most important areas of categorical data analysis within the social sciences and related fields.

The theory is illustrated by examples almost entirely of Danish origin. In particular I have drawn heavily on the Danish Welfare Study 1974, 1984. I am grateful to the project director Erik Jørgen Hansen for help and support over many years. Also the exercises are based on Danish data. As a help, solutions to theoretical problems and computer output for selected exercises are given in the Appendix. This should help readers without access to relevant statistical computer packages to work through the exercises.

The book requires a basic knowledge of statistical theory, but I have tried to keep mathematics at as elementary a level as possible.

References are not given in the main text, but collected in short sections called "Bibliographical notes" for the benefit of readers seeking additional reading.

Professor Leroy Folks has read the manuscript, corrected many errors and checked my language. I am very grateful for this invaluable help.

Any errors are of course entirely my responsibility.

Copenhagen, October 1996.   Erling B. Andersen

Contents

Chapter 1
Introduction 1
    1.1   The two-way table 1

Chapter 2
Basic Theory 6
    2.1   Introduction 6
    2.2   Exponential families 6
    2.3   Statistical inference in an exponential family ... 8
    2.4   The binomial distribution 12
    2.5   The multinomial distribution 13
    2.6   The Poisson distribution 20
    2.7   Composite hypotheses 21
    2.8   Applications to the multinomial distribution 25
    2.9   Log-linear models 29
    2.10 The two-way contingency table 31
    2.11 The numerical solution of the likelihood equations
           for the log-linear model 35
    2.12 Bibliographical notes 36
    2.13 Exercises 37

Chapter 3
Three-way contingency tables 42
    3.1   Log-linear models 42
    3.2   Log-linear hypotheses 46
    3.3   Estimation 49
    3.4   Testing hypotheses 59
    3.5   Interpretation of the log-linear parameters 65
    3.6   Choice of model 66
    3.7   Detection of model deviations 73
    3.8   Bibliographical notes 78
    3.9   Exercises 79

Chapter 4
Multi-dimensional contingency tables 84
    4.1   The log-linear model 84
    4.2   Classification and interpretation of log-linear models 87
    4.3   Choice of model 99
    4.4   Diagnostics 105
    4.5   Model search strategies 109
    4.6   Bibliographical notes 114
    4.7   Exercises 114

Chapter 5
Incomplete Tables ... 127
    5.1   Random and structural zeros ... 127
    5.2   Counting the number of degrees of freedom ... 129
    5.3   Validity of the χ^2-approximation ... 133
    5.4   Exercises ... 138

Chapter 6
The Logit Model ... 141
    6.1   The logit model ... 141
    6.2   Hypothesis testing in the logit model ... 144
    6.3   Logit models with higher order interactions ... 151
    6.4   The logit model as a regression model ... 154
    6.5   Bibliographical notes ... 155
    6.6   Exercises ... 156

Chapter 7
Logistic Regression Analysis ... 157
    7.1   The logistic regression model ... 157
    7.2   Estimation in the logistic regression model ... 159
    7.3   Numerical solution of the likelihood equations ... 162
    7.4   Checking the fit of the model ... 163
    7.5   Hypothesis testing ... 172
    7.6   Diagnostics ... 177
    7.7   Predictions ... 182
    7.8   Dummy variables ... 183
    7.9   Polytomous response variables ... 187
    7.10 Bibliographical notes ... 194
    7.11 Exercises ... 194

Chapter 8
Association Models ... 204
    8.1   Introduction ... 204
    8.2   Symmetry models ... 204
    8.3   Marginal homogeneity ... 209
    8.4   RC-association models ... 210
    8.5   Correspondence analysis ... 217
    8.6   Bibliographical notes ... 225
    8.5   Exercises ... 226

Appendix
Solutions and output to selected exercises. ... 235

References ... 260

Subject Index ... 264

Chapter 1

Introduction

1.1 The two-way table

A **two-way table**, or a **two-way contingency table,** gives the observed counts simultaneously for the categories of two categorical variables, as illustrated by the following example.

EXAMPLE 1.1. *In 1976 a random sample of approximately 5000 Danes was interviewed concerning a wide range of aspects of their private and working lives. We shall term this study: The Danish Welfare Study. From one of the reports from the Welfare Study (Hansen(1984)), we bring one of the tables concerning how often the interviewed attended meetings outside working hours and their social group. We shall meet the classification in social groups, used by the Danish National Institute for Social Research, several times in this book. Here is a short description of the five groups:*

Social group I	*Academics and main executives in the private and public sector.*
Social group II	*Second level executives in the private and public sector.*
Social group III	*Foremen, heads of sections etc. with less than 5 employees below them.*
Social group IV	*White collar workers and blue collar workers with special training.*
Social group V	*Blue collar workers.*

Table 1.1 now shows the cross-classification of the two categorical variables: "Frequency of attending meetings outside working hours" and "Social group" for the 1779 persons in the sample with age between 40 and 59.

TABLE 1.1. 1779 persons in Denmark between the age of 40 and 59 cross-classified according to the variables: Frequence of attending meetings and Social group.

Social group	Frequence of attending meetings					Total
	One or more times a week	One or more times a month	Approx. once every second month	A few times a year	Never	
I	17	27	13	24	25	106
II	25	57	17	49	55	203
III	38	91	41	217	213	600
IV	22	33	21	133	222	431
V	9	21	17	87	305	439
Total	111	229	109	510	820	1779

Source: Data from the Danish Welfare Study: Hansen (1984), appendix Table 14.

From elementary courses in statistics we know that the hypothesis of independence between two categorical variables, like those shown in Table 1.1, is tested by the Q-test statistic:

$$Q = \sum_i \sum_j \frac{(x_{ij} - m_{ij})^2}{m_{ij}} \tag{1.1}$$

where

$$m_{ij} = \frac{x_{i.} x_{.j}}{n}.$$

In these expressions x_{ij} is the observed count in cell ij (that is for the combination of category i for the row variable and category j for the column variable), m_{ij} is an estimate for the expected count in cell ij if the variables are independent, and n is the total number of counts in the table.

In the test statistic (1.1) the point is to compare the observed counts with the expected counts under independence, and to do so by only looking at one summarizing number, namely the observed number of Q. If Q is large we have reasons to believe that the hypothesis of independence is not true. If Q is small or moderate in value, we would tend to conclude that there is in fact independence. But let us return to Example 1.1 to see why a comparison between x_{ij} and m_{ij} is so important.

EXAMPLE 1.1 (continued). *Independence between variables "Frequence of attending meetings" and "Social group" obviously means that there is no difference* between the frequency with which persons in a low social group and persons in a high social group attend meetings outside working hours. If this is so, then the percentages over column categories in each row should be the same. Whether this is the case is illustrated by Table 1.2, where these percentages are calculated.

TABLE 1.2. Percentages rowwise for the data in Table I.1.

Frequence of attending meetings

Social group	One or more times a week	One or more times a month	Approx. once every second month	A few times a year	Never	Total
I	16.0	25.5	12.3	22.6	23.6	100.0
II	12.3	28.1	8.4	24.1	27.1	100.0
III	6.3	15.2	6.8	36.2	35.5	100.0
IV	5.1	7.7	4.9	30.9	51.5	100.0
V	2.1	4.8	3.9	19.8	69.5	100.0
Total	6.2	12.9	6.1	28.7	46.1	100.0

From these percentages is obvious that the independency hypothesis has to be rejected. Clearly the frequencies of attending meetings change with social group. Persons in a high social group are much more likely to attend meetings frequently than persons in a low social group. The observed value of Q is 262.7, which is a very high number. In our basic course in statistics, we learned to compare the observed value of Q with the percentiles of the χ^2-distribution with a number of degrees of freedom equal to the number of rows minus one times the number of columns minus one, here (5-1)(5-1) = 16. The 95%-percentile in a χ^2-distribution with 16 degrees of freedom is 26.3, so the Q-value we have observed is very unlikely, should the independency hypothesis hold.

When we perform a test based on the Q-test statistic, we are using a **statistical model** for the data. In case of Example 1.1 the essence of the test was to compare the observed counts x_{ij} with the expected counts m_{ij}. But the expected counts are just the common row totals for all rows, namely $x_{.1},...x_{.5}$ divided by n and then multiplied by the row marginals $x_{i.}$. Hence the expected counts are what we should expect in the rows, if the row percentages are equal. The failure of the independency hypothesis to describe the data, is thus equivalent to the fact, that an assumption of equal row percentages fails to fit the data in the table. A statistical model should, according to this line of reasoning, describe the distribution over column categories for each row and for the marginal at the bottom of the table, and compare these distributions.

The statistical distribution, which describes the distribution over a number of categories, given how many individuals are to be distributed, is (under certain important assumptions, which we leave out here) the **multinomial distribution**. It follows that the Q-test is a statistical test for comparing m multinomial distributions under the assumption that they are identical. In Example 1.1 we thus compare the multinomial distributions for the five rows of the table with the one for the column marginals. The multinomial distribution is for this - and other reasons - the central distribution for the models we shall study in this book.

Actually in real life applications it is more the exception than the rule that a hypothesis of independence in a two-way table is accepted. Hence we need statistical methods to describe why the hypothesis fails to describe the observed count. Obviously any study of how it can be, that our hypothesis of independence, or equivalently of equal row percentages, does not describe the observed counts in the two-way table, must be based on a study of the differences $x_{ij} - m_{ij}$.

It is rather common to see the square roots of the individual terms in the Q-test statistic, that is

$$\frac{x_{ij} - m_{ij}}{\sqrt{m_{ij}}},$$

as indicators of which cells in the two-way table contribute most to departures from independence. The problem with these indicators is that their relative expected magnitude under the hypothesis of independence varies with the number of rows and columns and with the values of the marginals. Hence we have no obvious "yardstick" for claiming when an indicator is large or small.

Fortunately it is known how to standardize the differences $x_{ij} - m_{ij}$ such that they approximately follow a standard normal distribution. The formula is as follows

$$\frac{x_{ij} - m_{ij}}{\sqrt{m_{ij}(1-x_{i.})(1-x_{.j})}}.$$

These standardized values are called **residuals** or, to emphasize that they are standardized, **standardized residuals**.

EXAMPLE 1.1. (continued) *Table 1.3 shows the indicators, where we have only divided with the square root of the expected counts, while Table 1.4 shows the standardized residuals.*

TABLE 1.3. Differences between observed and expected counts, divided by the square root of the expected count for the data in Table 1.1.

	Frequence of attending meetings				
Social group	One or more times a week	One or more times a month	Approx. once every second month	A few times a year	Never
I	4.04	3.62	2.55	-1.16	-3.41
II	3.47	6.04	1.29	-1.21	-3.99
III	0.09	1.57	0.70	3.43	-3.82
IV	-0.94	-3.02	-1.05	0.85	1.66
V	-3.51	-4.72	-1.91	-3.46	7.22

TABLE 1.4. Standardized residuals for the data in Table 1.1.

	Frequence of attending meetings				
Social group:	One or more times a week	One or more times a month	Approx. once every second month	A few times a year	Never
I	4.30	3.99	2.72	-1.41	-4.79
II	3.80	6.87	1.42	-1.52	-5.77
III	0.12	2.06	0.89	4.99	-6.39
IV	-1.12	-3.71	-1.25	1.16	2.59
V	-4.18	-5.83	-2.27	-4.73	11.32

From both tables we infer that the dependencies are such that the higher the social group, the more often people attend meetings outside working hours. Note that the numbers in Table 1.4 are markedly higher than in Table 1.3. For example cells 3.2, 4.5 and 5.3 show values numerically higher than the critical value 2, corresponding to the 97.5% percentile of the standard normal distribution, in Table 1.4, but not in Table 1.3. Thus in a situation with a few important significant departures from independence, we are likely to identify them from the standardized residuals; but we may miss them from a table like Table 1.3.

In subsequent chapters, we shall not only formulate models for contingency tables and develop tests for model check and for hypotheses regarding the parameters of the models, but also study methods for describing the nature of model departures and which parts of the data are the primary source for the model departures.

Chapter 2

Basic Theory

2.1 Introduction

Most of the models presented in this book belong to a class of models called **exponential families**. The statistical distributions, which we use most often, are the **multinomial distribution** and the **Poisson distribution**. Both these distributions in their basic form belong to the class of exponential families.

This chapter is, therefore, devoted to a treatment of the basic theory for models within the class of exponential families, especially models based on the multinomial distribution.

It is assumed that the reader is familiar with basic concepts in statistical distribution theory, estimation theory and the theory for testing statistical hypotheses. If this is not the case, it is advisable to consult an elementary text book in mathematical or theoretical statistics, for example Andersen, Jensen and Kousgård (1987).

2.2 Exponential families

Let $X_1,...,X_n$ be n independent, identically distributed random variables with common distribution, expressed through the **point probability**

$$P(X=x) = f(x|\theta) , \qquad (2.1)$$

which depends on the vector $\theta = (\theta_1,...,\theta_n)$ of real valued **parameters**.

Note: Vectors are in general not written with bold face type setting. With this notation, vectors can be recognized by not having subscripts, while individual elements have subscripts as in $\theta = (\theta_1,...,\theta_n)$.

The simultaneous distribution of $(X_1, ... ,X_n)$ is then expressed as the **joint point probability**

$$f(x_1,...,x_n|\theta) = \prod_{i=1}^{n} f(x_i|\theta) . \qquad (2.2)$$

If the point probability (2.1) has the form

$$\ln f(x|\theta) = \sum_{j=1}^{m} g_j(x)\phi_j(\theta) + h(x) - K(\theta) , \qquad (2.3)$$

where g_j and h are real valued functions of x, and ϕ_j and K are real valued functions of the vector θ, $f(x|\theta)$ is said to belong to an **exponential family**. The important part of expression (2.3) is that in logarithmic form the value of x and the value of the parameter vector θ only appear together as products $g_j(x)\phi_j(\theta)$ of certain real valued functions of x and θ.

Because of the constraint $\Sigma_x f(x|\theta) = 1$, the function $K(\theta)$ is implicitly given as a function of all the other functions through the relationship

$$K(\theta) = \ln\left\{\sum_x \exp\left(\sum_j g_j(x)\phi_j(\theta) + h(x)\right)\right\} , \qquad (2.4)$$

The smallest number m for which $\ln f(x|\theta)$ can be written in the form (2.3) is called the **dimension** of the exponential family.

If we introduce the **sufficient statistics**

$$t_j = \sum_{i=1}^{n} g_j(x_i) , \; j=1,...,m$$

and the **canonical parameters**

$$\tau_j = \phi_j(\theta) ,$$

Equations (2.2) and (2.3) can be written as

$$\ln f(x_1,...,x_n|\theta) = \ln f(x_1,...,x_n|\tau)$$
$$= \sum_{j=1}^{m} t_j\tau_j + \sum_{i=1}^{n} h(x_i) - nK(\tau) , \qquad (2.5)$$

where $\tau = (\tau_1,...,\tau_m)$. It is a consequence of (2.4), which can be written

$$K(\theta) = \ln\left\{\sum_x \exp\left(\sum_j g_j(x)\tau_j + h(x)\right)\right\} = K(\tau) ,$$

that $K(\theta)$ is a function of τ. In order not to unnecessarily complicate notation, we write $K(\tau)$, although K is strictly speaking not the same function of τ as $K(\theta)$ is of θ.

It is important, that the log-likelihood function (2.5) for an exponential family only depends on the observations through the values of the sufficient statistics t_j, and on the parameters through the values of the canonical parameters τ_j. It follows that for an exponential family one can only estimate parameters and test hypotheses for the canonical parameters, or set of parameters which are uniquely determined by the canonical parameters. It also follows that all information available in the data is summarized in the values of the sufficient statistics. Hence we only need the observed values of the sufficient statistics for statistical analyses. This means that we can concentrate on studying the canonical parameters, and that we can replace the log-likelihood function (2.5) with

$$\ln f(t|\tau) = \sum_j t_j \tau_j + h_1(t) - nK(\tau), \qquad (2.6)$$

where $t = (t_1,...,t_m)$ and

$$h_1(t) = \sum_{x_1,...,x_n|t} \left[\sum_i h(x_i) \right],$$

since

$$f(t|\tau) = \sum_{x_1,...,x_n|t} f(x_1,...,x_n|\tau).$$

2.3 Statistical inference in an exponential family

Since (2.6) is the logarithm of the joint point probability of the observed values of the sufficient statistics and all the information available concerning the canonical parameters is contained in the sufficient statistics, (2.6) is the log-likelihood function pertaining to statistical inference concerning the canonical parameters. We can thus write

$$\ln L(\tau) = \sum_j t_j \tau_j + h_1(t) - nK(\tau).$$

In order to maximize the log-likelihood function and thus obtain the **maximum likelihood estimates** (ML-estimates), we differentiate the log-likelihood function partially with respect to τ_j. The resulting **likelihood equations** become

BASIC THEORY

$$\frac{\partial \ln L(\tau)}{\partial \tau_j} = t_j - n\frac{\partial K(\tau)}{\partial \tau_j} = 0, \quad j=1,\ldots,m$$

or the m equations

$$t_j = n\frac{\partial K(\tau)}{\partial \tau_j}, \quad j=1,\ldots,m. \tag{2.7}$$

One of the advantages of working with distributions which belong to the exponential family is the simple form of the likelihood equations (2.7). A closer look at the properties of the function K make the likelihood equations even more attractive. From (2.3) and the condition $\Sigma_x f(x|\theta) = 1$ it follows that

$$\sum_x \exp\left\{\sum_j g_j(x)\tau_j + h(x) - K(\tau)\right\} = 1,$$

or

$$\exp\{K(\tau)\} = \sum_x \exp\left\{\sum_j g_j(x)\tau_j + h(x)\right\}.$$

Hence by differentiation with respect to τ_j we get

$$\exp\{K(\tau)\} \cdot \frac{\partial K(\tau)}{\partial \tau_j} = \sum_x g_j(x)\exp\left\{\sum_j g_j(x)\tau_j + h(x)\right\}.$$

If divided by $\exp\{K(\tau)\}$ the right hand side in this equation has the form

$$\sum_x g_j(x)f(x|\tau) = E[g_j(X)].$$

Hence

$$\frac{\partial K(\tau)}{\partial \tau_j} = E[g_j(X)]. \tag{2.8}$$

Summation then yields

$$E[T_j] = \sum_i E[g_j(X_i)] = n\frac{\partial K(\tau)}{\partial \tau_j}, \tag{2.9}$$

where T_j is the random variable corresponding to the sufficient statistic t_j.

If we compare (2.8) and (2.9) with (2.7) it becomes clear that the likelihood equations can also be written as

$$t_j = E[T_j], \quad j = 1,\ldots,m, \tag{2.10}$$

where, of course, the mean values $E[T_j]$ depend on the parameters.

Since for regular functions the partial derivatives of a function must be 0 at the point of maximum, Equations (2.7) and (2.10) show that it is a **necessary** condition for $\hat{\tau}_1,...,\hat{\tau}_n$ to be the ML-estimates that they satisfy Equations (2.10). When we are working with exponential families, it can also be established under which conditions it is a **sufficient condition** for obtaining the maximum, that the partial derivatives are 0, that is whether Equations (2.7) and (2.10) has a unique set of solutions, which are then the ML-estimates.

In order to formulate this important result, we need two new basic concepts: The **domain** and the **support**.

DEFINITION: *The domain D for an exponential family is a set within the range of variation for the τ's, for which the condition*

$$\sum_t \exp[t\tau' + h_1(t)] < \infty,$$

is satisfied.

Since it follows from (2.6) that f(t|τ) can be written

$$\frac{\exp[t\tau' + h_1(t)]}{\sum_t \exp[t\tau' + h_1(t)]} \qquad (2.11)$$

the domain consists of all those τ-vectors for which the likelihood function exists.

DEFINITION: *The support is the set T_0 of all t-vectors for which f(t|τ) is positive.*

From the form (2.11) of f it is clear that the support does not depend on τ.

After introducing the domain and the support we have the important result.

THEOREM 2.1. *If the domain is an open set in R^m there exists a unique solution to the likelihood equations*

$$t_j = E[T_j], j = 1,...,m.$$

if t is an interior point in the smallest convex set which includes all points of the support. This solution $\hat{\tau}$ is then the ML-estimate for τ.

In case the standard errors of the ML-estimates are also required, the following result can be used for suitable large sample sizes.

BASIC THEORY

THEOREM 2.2. *If τ is in the domain, $\hat{\tau}$ will converge in probability to τ and the asymptotic distribution of $\hat{\tau}$ satisfies*

$$\sqrt{n}(\hat{\tau}-\tau) \sim N_m(0, M^{-1}), \qquad (2.12)$$

where N_m is the m-dimensional normal distribution, 0 a vector of zero's and M a quadratic matrix with elements

$$m_{pq} = \frac{\partial^2 K(\tau)}{\partial \tau_p \partial \tau_q}.$$

In large samples we have accordingly

$$E[\hat{\tau}] = \tau \qquad (2.13)$$

and when var[$\hat{\tau}$] is the variance-covariance matrix

$$\text{var}[\hat{\tau}] = \frac{1}{n} \cdot M^{-1}. \qquad (2.14)$$

Confidence intervals for the canonical parameters with confidence level α can then be obtained as

$$\hat{\tau}_j \pm u_{1-\alpha/2} \sqrt{\frac{\hat{m}^{jj}}{n}},$$

where m^{jj} is the j'th diagonal element in M^{-1}, and $\hat{m}^{jj}$ is m^{jj} with the τ's replaced by their ML-estimates.

In order to test hypotheses of the form

$$H_0: \tau = \tau_0$$

or

$$H_0: \tau_j = \tau_{j0}, \, j = 1,\ldots,m,$$

the following result can be used, with the abbreviated notation $Q \sim \chi^2(m)$, when Q is χ^2-distributed with m degrees of freedom.

THEOREM 2.3. *If τ_0 is in the domain then the test statistic*

$$Z = -2\ln\left(\frac{L(\tau_0)}{L(\hat{\tau})}\right) \qquad (2.15)$$

is under H_0 asymptotically χ^2-distributed with m degrees of freedom, that is if $Q \sim \chi^2(m)$, then

$$P(Z \leq z) \to P(Q \leq z),$$

when $n \to \infty$.

Theorem 2.3 is used to test the hypothesis H_0 as follows. The level of significance p for testing H_0 based on the observed value z of the test statistic Z can be approximated by

$$p = P(Q \geq z) ,$$

where $Q \sim \chi^2(m)$. In order to test H_0 at a given level α, we reject H_0, if

$$z \geq \chi^2_{1-\alpha}(m) . \qquad (2.16)$$

2.4 The binomial distribution

Let $X_1,...,X_n$ be n independent binary random variables with possible values 0 and 1 and probability π of observing the value 1. The common point probability for the X's can then be written

$$f(x|\pi) = \pi^x(1-\pi)^{1-x} , \quad x = 1,0 .$$

The logarithm of $f(x|\pi)$ takes the form

$$\ln f(x|\pi) = x\ln\pi + (1-x)\ln(1-\pi) = x\ln\left(\frac{\pi}{1-\pi}\right) + \ln(1-\pi) .$$

$f(x|\pi)$ thus belongs to an exponential family with m=1, g(x)=x and canonical parameter

$$\tau = \ln\left(\frac{\pi}{1-\pi}\right) .$$

For n independent, identically distributed random variables with this distribution the sufficient statistic t is equal to the sum of the x's, or $t = \Sigma x_i$. But for n independent binary random variables with the same probability of x=1, the sum t is binomially distributed with number parameter n and probability parameter π. It follows that the binomial distribution for varying π belongs to the class of exponential families.

Since the theory for the binomial distribution is well known, we can check the general results in this very simple situation. The likelihood equation is

$$t = E[T] = n\pi \qquad (2.17)$$

so that the ML-estimate for π is t/n. In order to find the ML-estimate for the canonical parameter τ, we note that

$$\pi = \frac{\exp(\tau)}{1+\exp(\tau)}$$

such that the ML-estimate for τ is obtained by solving

BASIC THEORY 13

$$t = n\frac{\exp(\tau)}{1+\exp(\tau)}$$

or

$$\tau = \ln\left(\frac{t}{n-t}\right).$$

It is important to note, that while the ML-estimate for π can be calculated for all values of t this is not the case for τ where t=0 and t=n does not yield any estimates. The reason for this can be seen by looking at the **domain** and the **support**. The domain is the real line, since the point probability exists for all values of the canonical parameter τ. Note, however, that the extreme values 0 and 1 for π corresponds to τ being +∞ and -∞. This explains why there are ML-estimates for π for t=0 and 1, but not for τ. The same can be seen by looking at the support, which is the set of integers (0,1,...,n). According to Theorem 2.1 there is a unique solution to the likelihood equation if t is any point in the interior of the smallest interval containing the support, which is the interval [0,1]. But 0 and 1 are not interior points in this interval. Thus Theorem 2.1 does not say what happens for t=0 and t=n. There are good reasons for regarding t=0 and t=1 as extreme cases, which must be handled carefully. For example, note that the variance of T is 0 for $\pi = 0$ or 1. Any evaluation of standard errors for observed values of t=0 or 1 are, therefore, meaningless.

2.5 The multinomial distribution

Let $X_1,...,X_n$ be independent, identically distributed random variables, which can each attain the values 1,2,...,k with probabilities

$$P(X=j) = \pi_j , j = 1,...,k .$$

The simultaneous point probability of $X_1,...,X_n$ is then

$$f(x_1,...,x_n|\pi) = \pi_1^{t_1}...\pi_k^{t_k} ,$$

where

$$t_j = \text{number of x's equal to j} .$$

It follows that

$$\ln f(x_1,...,x_n|\pi) = \sum_j t_j \ln \pi_j ,$$

A comparison with Equation (2.5) then shows that the common distribution of the X's belongs to the class of exponential families with sufficient statistics $t_1, ... , t_k$ and canonical parameters $\tau_j = \ln \pi_j$, j=1,...,k..

To derive the exponential family directly from the point probability of X is possible, but requires a somewhat clumsy notation, cf. exercise 2.4.

The distribution of the vector $(T_1,...,T_k)$ of random variables is a multinomial distribution with number parameter n and probability parameters $\pi_1,...,\pi_k$, that is

$$f(t_1,...,t_k|\pi) = \binom{n}{t_1...t_k} \pi_1^{t_1} \cdot ... \cdot \pi_k^{t_k}$$

The log-likelihood function is therefore

$$\ln L(\pi) = \ln f(t_1,...,t_k|\pi) = \ln\binom{n}{t_1...t_k} + \sum_j t_j \ln \pi_j , \qquad (2.18)$$

From this equation one would perhaps conclude that the dimension of the exponential family is k. But that is not the case! The binomial distribution is a special case of the multinomial distribution for k=2. According to Example 2.1 the canonical parameter in the binomial distribution is $\tau = \ln[\pi/(1-\pi)]$. That the canonical parameters in the multinomial distribution seem to be just the $\ln\pi_j$'s should thus warn us that we have overlooked something. The thing we have actually overlooked are the linear constraints

$$\sum_{j=1}^{k} t_j = n \qquad (2.19)$$

and

$$\sum_{j=1}^{k} \pi_j = 1 . \qquad (2.20)$$

The constraint (2.19) implies that the log-likelihood function (2.18) can be written as

$$\ln L(\pi) = \text{const.} + \sum_{j=1}^{k-1} t_j \ln \pi_j + (n - \sum_{j=1}^{k-1} t_j)\ln \pi_k$$

$$= \text{const.} + \sum_{j=1}^{k-1} t_j (\ln \pi_j - \ln \pi_k) + n \cdot \ln \pi_k ,$$

where the term **const.** is a constant, which does not depend on the parameters. The log-likelihood function thus has the form

$$\ln L(\tau) = \text{const.} + \sum_{j=1}^{k-1} t_j \tau_j + n \cdot \ln \pi_k , \qquad (2.21)$$

showing that the multinomial distribution with no other constraints on the probability parameters than (2.20), is an exponential family of dimension at most m=k-1. In this unconstrained case the log-likelihood function can not be further reduced, and the correct dimension of the exponential family is m=k-1. The

BASIC THEORY

sufficient statistics are t_j, $j = 1,...,k - 1$ and the canonical parameters

$$\tau_j = \ln\pi_j - \ln\pi_k \ .$$

With a short notation we shall often write

$$(T_1,...,T_k) \sim M\ (n;\ \pi_1,...,\pi_k)$$

if $(T_1,...,T_k)$ has a multinomial distribution with number parameter n and probability parameters $\pi_1,...,\pi_k$.

In section 2.9 we shall discuss parametric multinomial distributions, where the canonical parameters τ_j are linear functions of a new set of parameters. In this case we call the model a log-linear multinomial model, or just a **log-linear model**. It follows from (2.21) that

$$K(\pi) = -\ln\pi_k \ .$$

K can also be expressed as a function af the canonical parameters, but we seldom need this form.

According to (2.10), the likelihood equations are

$$t_j = E[T_j]\ ,\ j = 1,...,k - 1 \ .$$

Since the marginal distribution of T_j is a binomial distribution, when the vector $(T_1,...,T_k)$ follows a multinomial distribution, the likelihood equations can also be written as

$$t_j = n\pi_j\ ,\ j = 1,...,k-1\ ,$$

with solutions

$$\hat{\pi}_j = \frac{t_j}{n}\ ,\ j = 1,...,k-1 \ . \tag{2.22}$$

It is an important and useful property of exponential families, that the likelihood equations often provide simple and direct estimates for parameters, which are not the canonical parameters but rather a reparametrization of the canonical parameters. In this connection a **reparametrization** means a set of parameters, which have a one-to-one relationship to the canonical parameters. Thus if ML-estimates for one set of parameters are obtained, the ML-estimates for the other set are obtained by using the one-to-one relationship between the parameters. The multinomial distribution is a simple - but typical - example of this property. If Equations (2.22) are satisfied for $j = 1,...,k-1$, then the constraints $\Sigma\pi_j = 1$ and $\Sigma t_j = n$ implies that (2.22) is also true for $j=k$, that is

$$\hat{\pi}_k = \frac{t_k}{n}\ ,$$

so that (2.22) can be replaced by

$$\hat{\pi}_j = \frac{t_j}{n}, \quad j=1,\ldots,k. \tag{2.23}$$

The ML-estimates for the canonical parameters are thus

$$\hat{\tau}_j = \ln\hat{\pi}_j - \ln\hat{\pi}_k = \ln t_j - \ln t_k. \tag{2.24}$$

As for the binomial distribution, where $t = 0$ and 1 are border line cases, $t_j = 0$ for any j is a border line case for the multinomial distribution, where there are no ML-estimates for the canonical parameters. Thus if $t_j = 0$ for any $j = 1, \ldots, k-1$ (and $t_k \neq 0$), then $\hat{\pi}_j = 0$, but τ_j will be $-\infty$, which does not belong to the domain. Also $t_k = 0$ will cause the τ's to be $+\infty$ or indetermined depending on the values of the other t's.

From Theorem 2.1 it can also be deduced that cases where one or more t's are zero are border line cases. Consider, for example, the case k=3. Here the sufficient statistics are t_1 and t_2. But since $t_1 + t_2 + t_3 = n$, t_1 and t_2 satisfy the inequalities

$$0 \leq t_1 + t_2 \leq n.$$

In the (t_1,t_2)-plane the convex extension of the support is accordingly the triangle shaded in Figure 2.1

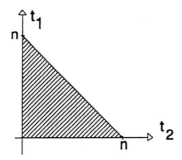

FIGURE 2.1. The convex extension of the support for the trinomial distribution.

Theorem 2.1 states that the likelihood equations pertaining to the canonical parameters have a unique set of solutions except perhaps on the boundary of the convex hull of the support. This means, in the present case, that unique solutions are guaranteed except along the edges of the triangle in Figure 2.1. But these edges are the t_2-axes with $t_1 = 0$, the t_1-axes with $t_2 = 0$ and the line $t_1 + t_2 = n$, where, because of (2.19), $t_3 = 0$. For all points in the interior of the triangle all three t's are positive and ML-estimates for the canonical parameters exists.

Consider now the hypothesis

Basic Theory

$$H_0: \pi_j = \pi_{j0}, j = 1,...,k. \tag{2.25}$$

In terms of the canonical parameters this hypothesis is

$$H_0: \tau_j = \ln\pi_{j0} - \ln\pi_{k0} = \tau_{j0}, j = 1,...,k-1. \tag{2.26}$$

Since the log-likelihood function according to (2.21) is equal to

$$\ln L(\tau) = \ln\binom{n}{t_1...t_k} + \sum_{j=1}^{k-1} t_j\tau_j + n\ln\pi_k,$$

the Z-test statistic in Theorem 2.3 in this case take the form

$$Z = -2\left(\sum_{j=1}^{k-1} T_j\tau_{j0} + n\ln\pi_{k0}\right) + 2\left(\sum_{j=1}^{k-1} T_j\hat{\tau}_j + n\ln\hat{\pi}_k\right).$$

In this formula it is not necessary to express π_k as a function of the τ's, since (2.26) together with (2.24) directly yields

$$Z = -2\sum_{j=1}^{k-1} T_j\left(\ln\pi_{j0} - \ln\pi_{k0}\right) - 2n\ln\pi_{k0} + 2\sum_{j=1}^{k-1} T_j\left(\ln T_j - \ln T_k\right) + 2n\ln(T_k/n)$$

$$= 2\sum_{j=1}^{k} T_j\ln T_j - 2\sum_{j=1}^{k} T_j\ln\pi_{j0} - 2n\ln(n),$$

or

$$Z = 2\sum_{j=1}^{k} T_j\left(\ln T_j - \ln(n\pi_{j0})\right). \tag{2.27}$$

The test statistic for the hypothesis (2.25) of known probability parameters in the multinomial distribution thus has the form

$$Z = 2\sum \text{observed }(\ln(\text{observed}) - \ln(\text{expected})),$$

We shall meet this form, which is characteristic for the test statistics, many times in this book.

It follows from Theorem 2.3 that

$$Z \sim \chi^2(k-1),$$

such that we reject H_0 at approximate level α, if the observed value of Z satisfies

$$z \geq \chi_{1-\alpha}^2(k-1).$$

Some readers may object, that the detour around the canonical parameters, when formulating the hypothesis as (2.26) and testing it by means of (2.27), is unnecessary, since both the hypothesis and the resulting test statistic only depend on the sufficient statistics and the π_{j0}'s. But the formulation in terms of the canonical parameters helps us to keep in mind:

(i) The hypothesis (2.25) only concerns k-1 unconstrained parameters, because one of the parameters, for example π_k, is a function of the other π's.

(ii) The ML-estimates for the canonical parameters does not exist if any t_j is 0. This corresponds to cases, where a term in the test statistic Z is 0, invalidating the approximation to the χ^2-distribution. Also cases where π_{j0} is close to 0 can cause the approximation of Z to a χ^2-distribution to be dubious.

EXAMPLE 2.1. *Traffic accidents.*
Table 2.1 shows the distribution of traffic accidents involving pedestrians in Denmark in 1981 by days of the week.

TABLE 2.1. Distribution of traffic accidents involving pedestrians in Denmark in 1981 by days of the week.

Weekday	Monday	Tuesday	Wednesday	Thursday	Friday	Saturday	Sunday	Total
Number of accidents	279	256	230	304	330	210	130	1739

Source: Road traffic accidents. Publication 1982:8. Statistics Denmark. Table 5.3

If we assume that the accidents happen independently of each other, we have 1739 observations of accidents, which all can fall on any day of the week. Hence the observed vector $(t_1,...,t_7)$ of accidents on the seven days follows a multinomial distribution. We want to test the hypothesis that a randomly selected accident is equally likely to fall on any of the 7 days. This is also the hypothesis of an equal distribution of accidents over the seven days of a week. In terms of the parameters of the multinomial distribution, we have $\pi_j = \pi_{j0} = 1/7$ for all j=1,...,7, if all 7 π's are equal. The expected numbers under the hypothesis are then

$$n\pi_{j0} = \frac{1739}{7} = 248.43 .$$

With the values in Table 2.1, the observed value of (2.27) is easily computed to be z = 115.83. In this case $Z \sim \chi^2(6)$, such that at level 0.05 we must reject the hypothesis if $z > \chi^2_{0.95}(6) = 12.6$. Hence we reject the hypothesis of accidents happening equally likely on all seven days of the week.

BASIC THEORY 19

EXAMPLE 2.2. *Representativity.*
Table 2.2 shows the number of girls between the age of 7 and the age of 15, who were selected for interviews in connection with a study by the Danish National Institute of Social Research regarding school children's use of their time outside school hours.

TABLE 2.2. The number of girls between the age of 7 and the age of 15, who were selected for interviews

Age	7	8	9	10	11	12	13	14	15	Total
Number interviewed	40	28	38	30	38	45	39	34	48	340

Source: Children's leisure time. Report 95.2. The Danish National Institute of Social Research. Appendix Table 1.

In the report on Children's leisure time the percentage of Danish girls on January 1st 1993 was also shown. These percentages are shown as frequencies in Table 2.3 together with the expected numbers in the age categories if the sample of girls is representative for the Danish population of girls in the given age categories. The sample is representative if a girl is selected in the sample with a probability π_{j0} equal to the relative size of the female population in her age interval j. The expected numbers are then $n\pi_{j0}$, $j=1,...,7$.

TABLE 2.3. The frequency of girls between age 7 and age 15 in Denmark on January 1st 1993 and the expected numbers in a representative sample of size 340.

Age	7	8	9	10	11	12	13	14	15	Total
Frequency 1/1 1993	.107	.103	.102	.105	.106	.114	.117	.123	.123	340
Expected numbers	36.4	35.0	34.7	35.7	36.0	38.8	39.8	41.8	41.8	340.0

With the values in Table 2.2 and Table 2.4 the observed value of (2.27) is z = 6.64. In this case $Z \sim \chi^2(8)$, such that at level 0.05 we must reject the hypothesis if $z > \chi^2_{0.95}(8) = 15.5$. Hence we accept the hypothesis and can conclude that the sample of girls is, in fact, representative of the population of 7 to 15 year old girls in Denmark.

2.6 The Poisson distribution

The Poisson distribution also belongs to the class of exponential families. Let $X_1,...,X_n$ be n independent random variables with common Poisson point probability

$$f(x|\lambda) = \frac{\lambda^x}{x!} e^{-\lambda}.$$

In logarithmic form, we get

$$\ln f(x|\lambda) = \ln(x!) + x\ln\lambda - \lambda,$$

showing that the sufficient statistic is $t = \Sigma x_i$ and the canonical parameter $\tau = \ln\lambda$. It then follows from the general result, Theorem 2.1, that the likelihood equation for estimating λ or τ is $t = E[T]$. Due to the addition rule for the Poisson distribution T is also Poisson distributed with parameter $n\lambda$. The ML-estimate for λ is accordingly obtained by solving $t = n\lambda$, or

$$\hat{\lambda} = \frac{t}{n} = \bar{x}.$$

The support for the Poisson distribution is the set of all non-negative integers. Hence the likelihood equation according to Theorem 2.1 has a guaranteed unique solution for all values of t in the open interval $(0,+\infty)$. For the Poisson distribution the border line case is thus $t=0$, that is the case where all x's are 0. In this case the canonical parameter τ has estimated value $-\infty$ and, even though the ML-estimate (with the value 0) for λ formally exists, $\text{var}[T]=0$ for $\lambda=0$, so we have to be careful here.

There is an important connection between the Poisson distribution and the multinomial distribution, which is useful in some situations. Let $X_1,...,X_k$ be independently Poisson distributed random variables with parameters $\lambda_1,...,\lambda_k$. The conditional distribution of $X_1,...,X_k$ given the value t of the sum $T = \Sigma X_i$ is then a multinomial distribution with number parameter $n=t$ and probability parameters $\pi_j = \lambda_j / \lambda_.$, where $\lambda_.$ is the sum $\Sigma\lambda_j$ of the λ's. In formula form we have

$$(X_1, ... ,X_n | T=n) \sim M (n; \lambda_1/\lambda_., ... ,\lambda_k/\lambda_.) \tag{2.28}$$

The proof of (2.28) is straightforward and is left to the reader, cf. exercise 2.3.

This connection between the Poisson distribution and the multinomial distribution allows us to treat many situations, where the "correct" model would seem to be a Poisson distribution, by the statistical methods developed in this book, which are in almost all cases based on the multinomial distribution or the binomial distribution. One typical example is Example 2.1. Many would argue that the "correct" model would be to assume that accidents follow a Poisson process. That is the probability that an accident happens in the time interval (t,t+dt) on a Monday is approximately $\lambda_1 dt$ for small values of dt. In this case λ_1 is the intensity of accidents per time unit on Mondays. The number of accidents x_1 on Mondays will then be Poisson distributed with mean value $\lambda_1 = (365/7)$. But according to (2.28) we can treat the vector $(x_1,...,x_7)$ of observed accidents for all seven days of the week as multinomially distributed with parameters $\pi_j = \lambda_j / \lambda_.$, $j = 1, ... 7$.

2.7 Composite hypotheses

In many cases the relevant hypothesis to test involves only some of the canonical parameters. The relevant hypotheses may for example be

$$H_0: \tau_j = \tau_{j0}, j = 1,...,r$$

with $r < m$.

Under H_0 there are in this case m-r parameters, which are not specified and accordingly must be estimated. Under H_0 the log-likelihood function (2.6) has the form

$$\ln L(\tau) = \sum_{j=1}^{r} t_j \tau_{j0} + \sum_{j=r+1}^{m} t_j \tau_j + h_1(t) - nK(\tau),$$

showing that the model still belongs to an exponential family, but now of dimension m-r. The likelihood equations under H_0 therefore become

$$t_j = E[T_j], j = r+1,...,m. \qquad (2.29)$$

If we denote the solutions to (2.29) $\tilde{\tau}_{r+1},...,\tilde{\tau}_m$, the value of the log-likelihood function becomes

$$\ln L(\tilde{\tau}) = \sum_{j=1}^{r} t_j \tau_{j0} + \sum_{j=r+1}^{m} t_j \tilde{\tau}_j + h_1(t) - nK(\tilde{\tau}),$$

where $\tilde{\tau} = (\tau_{10},...,\tau_{r0}, \tilde{\tau}_{r+1},...,\tilde{\tau}_m)$.

In order to test H_0 we can, therefore, use the test quantity

$$Z(H_0) = -2\ln\frac{L(\tilde{\tau})}{L(\hat{\tau})}, \qquad (2.30)$$

where $\hat{\tau}$ is the vector of unconstrained ML-estimates, that is τ estimated without H_0.

The asymptotic distribution of of $Z(H_0)$ follows from the following theorem:

THEOREM 2.4. *Let the vector*

$$\tau = (\tau_{10},...,\tau_{r0},\tau_{r+1},...,\tau_m),$$

be in the domain, where $\tau_{r+1},...,\tau_m$ *are the true, unknown values of the τ's which are not specified under the hypothesis. Then $Z(H_0)$ is under H_0 asymptotically χ^2-distributed with r degrees of freedom, i.e. for any observed value z of $Z(H_0)$*

$$P(Z(H_0) \leq z) \to P(Q \leq z),$$

when $n \to \infty$*, where* $Q \sim \chi^2(r)$.

The hypothesis H_0 can accordingly be rejected at approximate level α, if the observed value z of $Z(H_0)$ satisfies

$$z \geq \chi^2_{1-\alpha}(r) ,$$

or by evaluating the relative size of the level of significance $p = P(Q \geq z)$ for the observed value z of $Z(H_0)$, where approximately $Q \sim \chi^2(r)$.

We shall often write in the abbreviated form

$$Z(H_0) \sim \chi^2(r)$$

if $Z(H_0)$ has an approximate χ^2 - distribution with df degrees of freedom.

Theorem 2.4 can be extended to cases with **constrained canonical parameters**. We shall say that a hypothesis H_0 concerns constrained canonical parameters, if two conditions are satisfied.

(i) $$H_0 : \psi_j(\tau_1,...,\tau_m) = \psi_{j0} , \; j = 1,...,r ,\qquad(2.31)$$
and
(ii) $$\psi_j = \psi_j(\tau_1,...,\tau_m) , \; j = 1,...,m \qquad(2.32)$$

is a **reparametrization** of the model in the sense that there is a one to one correspondence between the ψ's and the τ's determined by the Equations (2.32). In this situation let $\hat{\tau}$ be the vector of ML-estimates for the τ's under the constraints (2.31) and $\hat{\tau}$ the ML-estimates for the unconstrained τ's in the test statistic (2.30). We then have the following more general form of Theorem 2.4.

THEOREM 2.5. *Assume the vector τ of canonical parameters obtained by solving (2.32) with the ψ-vector given by*

$$\psi = (\psi_{10},...,\psi_{r0},\psi_{r+1},...,\psi_m) ,$$

is in the domain. Then under H_0, given by (2.31), $Z(H_0)$ is asymptotically χ^2-distributed with r degrees of freedom, i.e.

$$P(Z(H_0) \leq z) \to P(Q \leq z) ,$$

when $n \to \infty$, where $Q \sim \chi^2(r)$.

Typical examples of hypotheses expressed through constraints on the canonical parameters are as follows:

(a) $$H_0: \tau_1 +...+\tau_m = 0 .$$

Here the ψ's are given by

BASIC THEORY

$$\psi_1 = \tau_1 + ... + \tau_m ,$$
$$\psi_2 = \tau_2 ,$$
$$...$$
$$\psi_m = \tau_m .$$

Then $H_0 : \psi_1 = 0$, with all the remaining ψ's unspecified.

(b) $\qquad\qquad H_0 : \tau_1 = ... = \tau_m .$

Here the ψ's are given by

$$\psi_1 = \tau_1 - \tau_m ,$$
$$...$$
$$\psi_{m-1} = \tau_{m-1} - \tau_m$$

and

$$\psi_m = \tau_m .$$

Then $H_0 : \psi_j = 0$, $j=1,...,m-1$.

Another important result is obtained if we only let the vector of parameters include some of the τ's, for example by putting $\tau_{s+1} = ... = \tau_m = 0$, with $r < s$. Then under the hypothesis

$$H_0: \tau_j = \tau_{j0} , j = 1,...,r ,$$

the ML-estimates for the canonical parameters will be

$$\tilde{\tau} = (\tau_{10},...,\tau_{r0},\tilde{\tau}_{r+1},...,\tilde{\tau}_s,0,...,0) .$$

Without the hypothesis, the ML-estimates will be

$$\hat{\tau} = (\hat{\tau}_1,...,\hat{\tau}_r,\hat{\tau}_{r+1},...,\hat{\tau}_s,0,...,0) .$$

Consider now the hypothesis

$$H_0^*: \tau_j = 0 , j = s+1,...,m .$$

Then $\tilde{\tau}$ are the ML-estimates under the hypothesis that both H_0 or H_0^* holds, while $\hat{\tau}$ is the vector of ML-estimates under H_0^*.

It makes the notation clearer if we write the likelihood function $L(\tau)$ as $L(\tilde{\tau}|H_0)$ when $\tilde{\tau}$ is estimated under H_0. The likelihood ratio test of H_0 against the alternative that the hypothesis H_0^* hold (rather than against the model with unspecified canonical parameters) can with this notation be based on the test statistic

$$Z(H_0|H_0^*) = -2\ln \frac{L(\hat{\tau}|H_0,H_0^*)}{L(\hat{\tau}|H_0^*)}$$

or

$$Z(H_0|H_0^*) = -2\ln L(\hat{\tau}|H_0,H_0^*) + 2\ln L(\hat{\tau}|H_0^*) . \qquad (2.33)$$

But since $H_0 \subset H_0^*$ in the sense that if H_0 is satisfied then also H_0^* is satisfied, we can rewrite (2.33) as

$$Z(H_0|H_0^*) = -2\ln L(\hat{\tau}|H_0) + 2\ln L(\hat{\tau}|H_0^*) . \qquad (2.34)$$

Here $L(\hat{\tau}|H_0)$, in accordance with our notation, is the likelihood function with the unconstrained parameters estimated under both H_0 and H_0^*, while $L(\hat{\tau}|H_0^*)$ is the likelihood function with only the unconstrained parameters under H_0^* estimated.

Theorem 2.4 is also valid for this situation because we are just testing r specified values of canonical parameters in an exponential family with s (rather than m) unconstrained canonical parameters $\tau_1,...,\tau_s$. The asymptotic distribution of $Z(H_0|H_0^*)$ is therefore, as in Theorem 2.4, a χ^2-distribution with r degrees of freedom. In summary we have.

THEOREM 2.6. *Let the vector*

$$\tau = (\tau_{10},...,\tau_{r0},\tau_{r+1},...,\tau_s) ,$$

be in the domain, where $\tau_{r+1},...,\tau_s$ are the true, unknown values of the τ's, which are unspecified under the hypothesis. Then under H_0 the test statistic $Z(H_0|H_0^*)$ given by (2.34) is asymptotically χ^2-distributed with r degrees of freedom, i.e

$$P(Z(H_0|H_0^*) \leq z) \rightarrow P(Q \leq z) ,$$

when $n \rightarrow \infty$, where $Q \sim \chi^2(r)$.

For later use we rewrite the test statistic (2.34). Noting that

$$Z(H_0|H_0^*) = (-2\ln L(\hat{\tau}|H_0) + 2\ln L(\hat{\tau})) + (2\ln L(\hat{\tau}|H_0^*) - 2\ln L(\hat{\tau})) ,$$

where $L(\hat{\tau})$ is the likelihood with τ estimated unconstrained, we get

$$Z(H_0|H_0^*) = Z(H_0) - Z(H_0^*) \qquad (2.35)$$

We shall use the abbreviated notation

$$Z(H_0|H_0^*) \sim \chi^2(r) ,$$

for the result in Theorem 2.6.

Theorems 2.4 and 2.5 thus show that $Z(H_0) \sim \chi^2(df)$, where df can either be calculated as the number of constrained canonical parameters, or as the difference

BASIC THEORY 25

between the total number of unconstrained canonical parameters and the number specified under the hypothesis.

Theorem 2.6 on the other hand establishes that if we test H_0 given that the hypothesis H_0^* is already satisfied, then $Z(H_0) - Z(H_0^*) \sim \chi^2(df)$, where df is the number of canonical parameters specified under H_0, or alternatively the difference between the number to be estimated under H_0 and the number to be estimated under H_0^*.

2.8 Applications to the multinomial distribution

In section 2.5 we derived the test statistic (2.27) for the hypothesis of all the probability parameters having specified values. We now consider typical cases, where under the hypothesis certain restrictions are imposed on the π's.

Matters are much simplified for the multinomial distribution when we recall that $\pi_1,...,\pi_{k-1}$ is a reparametrization of the canonical parameters $\tau_1,...,\tau_{k-1}$ and that $-2 \ln L$ can be expressed both as

$$-2\ln L(\tau) = \text{const.} - 2\sum_{j=1}^{k-1} t_j \tau_j - 2n\ln \pi_k \qquad (2.36)$$

and as

$$-2\ln L(\tau) = \text{const.} - 2\sum_{j=1}^{k} t_j \ln \pi_j , \qquad (2.37)$$

where in (2.37) the π's are functions of the τ's. It follows that if we can determine the ML-estimates for the π's under any composite hypothesis imposing constraints on the π's, then the test statistic $Z(H_0)$ in (2.30) or the test statistic $Z(H_0|H_0^*)$ in (2.34) can be expressed directly in terms of these π-estimates. As examples we consider the following two important and often met cases.

(a) $$\pi_1 = ... = \pi_r \quad , \quad \pi_{r+1} = ... = \pi_k .$$

In this case the log-likelihood function according to (2.36) is

$$\ln L = \text{const.} + (\sum_{j=1}^{r} t_j)\tau_r + n\ln \pi_k , \qquad (2.38)$$

since $\tau_j = 0$ for $j=r+1,...,k-1$, when $\pi_j = \pi_k$ for k>r. The likelihood equation for estimating the common value τ_r of the first r τ's under H_0 is, therefore,

$$t^{(r)} = \sum_{j=1}^{r} t_j = \sum_{j=1}^{r} E[T_j] = \sum_{j=1}^{r} n\pi_j = nr\pi_r,$$

with solution

$$\hat{\pi}_j = \hat{\pi}_r = \frac{t^{(r)}}{rn}, \; j = 1,\ldots,r.$$

From $\Sigma \pi_j = 1$ and $\pi_{r+1} = \ldots = \pi_k$ it then follows that

$$\hat{\pi}_j = \frac{t^{(k-r)}}{n(k-r)}, \; j = r+1,\ldots,k,$$

where

$$t^{(k-r)} = n - t^{(r)} = \sum_{j=r+1}^{k} t_j.$$

Without H_0 the ML-estimates are

$$\hat{\pi}_j = \frac{t_j}{n}.$$

Equations (2.30) and (2.37) therefore yield

$$Z(H_0) = -2\ln L(\hat{\hat{\tau}}) + 2\ln L(\hat{\tau})$$

$$= 2\sum_{j=1}^{k} T_j \left(\ln \frac{T_j}{n} - \ln \hat{\hat{\pi}}_j \right).$$

or

$$Z(H_0) = 2\sum_{j=1}^{k} T_j \left[\ln T_j - \ln(n\hat{\hat{\pi}}_j) \right]. \qquad (2.39)$$

The test statistic thus again has the typical form

$$Z = 2 \Sigma \text{ observed } (\ln(\text{observed}) - \ln(\text{expected}))$$

which we met in Section 2.5, Equation (2.27).

The number of degrees of freedom for $Z(H_0)$ can be found by noting that only one parameter has to be estimated according to (2.38). Thus the number of constrained parameters is k-1-1=k-2. We can also count the number of τ's, which are specified under H_0. Since $\tau_j=0$ for j=r+1,...,k-1 and the first r τ's are equal, (k-1-r)+(r-1) = k-2 canonical parameters are specified. The distribution of the test statistic $Z(H_0)$ is thus approximately

… BASIC THEORY

$$Z(H_0) \sim \chi^2(k-2),$$

and we reject H_0 at approximate level of significance α, if the observed value z of $Z(H_0)$ satisfies

$$z > \chi^2_{1-\alpha}(k-2).$$

The next case is

(b) $\quad\quad \pi_1 = \ldots = \pi_r \,,\quad \pi_j \text{ unspecified}, j = r+1,\ldots,k.$

In this case the log-likelihood function is

$$\ln L(\tau) = \text{const.} + \left(\sum_{j=1}^{r} t_j\right)\tau_r + \sum_{j=r+1}^{k-1} t_j\tau_j + n\ln\pi_k,$$

since τ_j is unspecified for $j = r+1,\ldots,k-1$ when π_j is unspecified for $k>r$. The likelihood equation for estimating the common value τ_r of the first r τ's under H_0 is

$$t^{(r)} = \sum_{j=1}^{r} t_j = \sum_{j=1}^{r} E[T_j] = \sum_{j=1}^{r} n\pi_r,$$

with solution

$$\hat{\pi}_j = \hat{\pi}_r = \frac{t^{(r)}}{rn}, \, j = 1,\ldots,r.$$

For $j=1,\ldots,k-1$ we get the likelihood equations $t_j = n\pi_j$ with solutions

$$\hat{\pi}_j = \frac{t_j}{n}, \, j = r+1,\ldots,k-1.$$

It is easy to see that these estimates, due to $\Sigma\pi = 1$, imply that

$$\hat{\pi}_k = \frac{t_k}{n}.$$

Without H_0 the ML-estimates are still

$$\hat{\pi}_j = \frac{t_j}{n}.$$

Equations (2.30) and (2.37) therefore again yield

$$Z(H_0) = 2\sum_{j=1}^{k} T_j\left[\ln T_j - \ln(n\hat{\pi}_j)\right]. \quad\quad (2.40)$$

Also in this case the test statistic has the form

$$Z = 2\,\Sigma \text{ observed }(\ln(\text{observed}) - \ln(\text{expected})).$$

The degrees of freedom for $Z(H_0)$ in this case can most easily be determined by noting that r-1 canonical parameters are specified under H_0. The distribution of the test statistic $Z(H_0)$ is thus approximately

$$Z(H_0) \sim \chi^2(r-1).$$

For case (b) we accordingly reject H_0 at approximate level of significance α if the observed value z of $Z(H_0)$ satisfies

$$z > \chi^2_{1-\alpha}(k-1) \, .$$

EXAMPLE 2.1 (continued). *It is obvious from Table 2.1 that the number of accidents is lower in the weekend. A tempting hypothesis is, therefore,*

$$H_0 : \pi_1 = ... = \pi_5 \, , \pi_6 = \pi_7.$$

This is case (a) with k=7 and r=5. The common estimate of π_1 to π_5 is

$$\hat{\pi}_5 = \frac{279+256+230+304+330}{5 \cdot 1739} = 0.1609 \, .$$

The common expected number of accidents on Mondays to Fridays is thus
$$0.1609 \cdot 1739 = 279.8 \, .$$

In the same way the common estimate of π_6 and π_7 is

$$\hat{\pi}_6 = \hat{\pi}_7 = \frac{210+130}{2 \cdot 1739} = 0.0978 \, ,$$

The common expected number of accidents on week-ends is thus
$$0.0978 \cdot 1739 = 170.0 \, .$$

With these expected values under the hypothesis, the observed value of $Z(H_0)$ is z=41.08. With k=7 the number of degrees of freedom for the approximating χ^2-distribution is 5. Since $P(Q>41.08)= 0.0000001$ if $Q \sim \chi^2(5)$ we can not accept the hypothesis.

As a final attempt, we could try to test whether the accidents are equally distributed over the first five weekdays, lower in the week-end, but not necessarily equally distributed between Saturday and Sunday. This is case (b), again with k=7 and r=5. The expected numbers for Monday to Friday are the same as for case (a), but for Saturday and Sunday, the observed and expected values are now equal. In this case the value of $Z(H_0)$ get the observed value z=22.08 and the number of degrees of freedom is 5-1=4. But since $P(Q>22.08)= 0.0002$ if $Q \sim \chi^2(4)$, we must still reject the hypothesis. The general conclusion is, therefore, that there is a clear variation of accidents over days of the week. It would be an error, from an applied statistician's point of view, to try to test the very specified hypothesis, that the π's are equal for Monday, Tuesday and Wednesday, then change to a higher level on

BASIC THEORY   29

Thursday and Friday, and finally decrease over the week-end. If we do so, we would use our data to formulate a hypothesis rather than to test a hypothesis.

2.9 Log-linear models

First we give two important definitions. First, we shall call a multinomial model the **saturated model** if the canonical parameters $\tau_j = \ln\pi_j - \ln\pi_k$ are unconstrained. Second we shall call a multinomial model a **log-linear model** if the canonical parameters in the saturated model are linear functions of a new set of parameters $(\theta_1,...,\theta_m)$, i.e

$$\tau_j = \ln\pi_j - \ln\pi_k = \sum_p w_{jp}\theta_p \, , \, j = 1,...,k-1 \qquad (2.41)$$

where the w's are known constants and θ_p, p = 1,...,m are the real parameters of the model. The matrix **W** with elements w_{jp}, j=1,...,k-1, p=1,...,m is called the **design matrix**.

Since there is a one to one correspondence between the canonical parameters and the probability parameters, the π's can be expressed in terms of the θ's, namely as

$$\pi_j = \frac{\exp\left(\sum_p w_{jp}\theta_p\right)}{\sum_{q=1}^{k} \exp\left(\sum_p w_{qp}\theta_p\right)} . \qquad (2.42)$$

From (2.41) follows that we can put $w_{kp} = 0$, p = 1,...,m, which means that (2.42) can be written as

$$\pi_j = \frac{\exp\left(\sum_p w_{jp}\theta_p\right)}{1 + \sum_{q=1}^{k-1} \exp\left(\sum_p w_{qp}\theta_p\right)} . \qquad (2.43)$$

If $(T_1,...,T_k)$ follows a multinomial distribution the log-likelihood function for observed values $(t_1,...,t_k)$ is

$$\ln L(\tau) = \text{const.} + \sum_{j=1}^{k-1} t_j\tau_j + n\ln\pi_k .$$

Hence under a log-linear model lnL has the form

$$\ln L(\theta) = \text{const.} + \sum_p \theta_p \sum_j w_{jp}t_j + n\ln\pi_k .$$

according to (2.41). The model thus still belongs to the class of exponential families, but the canonical parameters are now the θ's and the sufficient statistics are

$$t_p^* = \sum_j w_{jp} t_j \ , \ p=1,...,m \ .$$

It follows that the ML-estimators are obtained as solutions to the equations

$$t_p^* = E[T_p^*] = \sum_j w_{jp} E[T_j] = \sum_j w_{jp}(n\pi_j) \ , \ p=1,...,m \ .$$

or as

$$\sum_j w_{jp} t_j = \sum_j w_{jp}(n\pi_j) \ , \ p=1,...,m \ . \qquad (2.44)$$

The standard errors of the ML-estimators are obtained from an application of Theorem 2.2. It is rather straightforward (although we omit the details) to show that the matrix **M** has elements

$$m_{pq} = \left[\sum_j w_{jp} w_{jq} \pi_j - \sum_j w_{jp} \pi_j \sum_s w_{sq} \pi_s \right],$$

where to save space π_j is a function of θ given by (2.43). If we introduce the matrix **V** of dimension k×k with elements

$$v_{jj} = \pi_j(1-\pi_j) \ ,$$

$$v_{js} = -\pi_j \pi_s \ , \ j \neq s \ ,$$

and extend the matrix **W** of dimension (k-1)×m to a matrix of dimension k×m by adding a last line with elements $w_{kp} = 0$, p=1,...m, **M** can be written

$$\mathbf{M} = \mathbf{W'VW}. \qquad (2.45)$$

From Theorem 2.2 it then follows that

$$\text{var}[\hat{\theta}_p] = \frac{1}{n} m^{pp} \ ,$$

where m^{pq} is the (p,q)'th element in the inverse matrix $\mathbf{M}^{-1}$ of **M**.

The matrix **M** = **W'VW** is a key element also in connection with analyses of residuals. In fact it can be shown that

$$\text{var}[T_j - n\hat{\pi}_j] = n\pi_j(1-\pi_j)(1-h_{jj}) \ , \qquad (2.46)$$

where h_{jj} is the j'th diagonal element in the matrix

$$\mathbf{H} = (\mathbf{V}^{½} \mathbf{W} \mathbf{M}^{-1} \mathbf{W'} \mathbf{V}^{½})^{-1} \ , \qquad (2.47)$$

BASIC THEORY

$V^{1/2}$ being diagonal with diagonal elements

$$\sqrt{\pi_j \cdot (1-\pi_j)} \, .$$

H is called the "hat" matrix.

It is thus relatively easy to compute the necessary estimates and test statistics for log-linear models. Admittedly it can not be done be hand, and hardly by pocket computers. But it is easy to program the computations for a computer, and for many important special cases elements of the hat matrix are included in standard statistical computer packages.

2.10 The two-way contingency table

In Chapter 1 we discussed the two-way contingency table in an introductory way. The two-way table is the most basic of all log-linear models and also the model, which - in its multivariate version - is the main topic of this book.

To be in accordance with the notation we shall use later in the book, we shall denote the observed numbers in the I×J cells of a two-way table x_{ij}. The model we shall use is a multinomial model over the I×J cells, that is

$$(X_{11},...,X_{IJ}) \sim M(n; \pi_{11},...,\pi_{IJ})$$

The independence hypothesis states that

$$H_0 : \pi_{ij} = \pi_{i.} \pi_{.j} \, ,$$

where $\pi_{i.} = \Sigma_j \pi_{ij}$ and $\pi_{.j} = \Sigma_i \pi_{ij}$

The canonical parameters for the multinomial distribution over the cells are

$$\tau_{ij} = \ln \pi_{ij} - \ln \pi_{IJ} \, .$$

Hence under H_0, we have the constraints

$$\tau_{ij} = \ln \pi_{ij} - \ln \pi_{IJ} = (\ln \pi_{i.} - \ln \pi_{I.}) + (\ln \pi_{.j} - \ln \pi_{.J}) \quad (2.48)$$

on the canonical parameters. This is a log-linear model with I+J-2 new parameters $\theta_1,...,\theta_{I+J-2}$, since τ_{ij} according to (2.48) can be written

$$\tau_{ij} = \theta_i + \theta_{I-1+j} \, , \quad (2.49)$$

where

$$\theta_i = \ln \pi_{i.} - \ln \pi_{I.} \, , \, i = 1,...,I-1$$

and
$$\theta_{I-1+j} = \ln\pi_{\cdot j} - \ln\pi_{\cdot J}, j = 1,...J-1.$$

Equation (2.49) shows that the two-way contingency table under the independence hypothesis is a log-linear model with all weights w being either 0 or 1. To illustrate how the w's look in a typical situation, we take as an example the 3×3 table, where the w's are given by Table 2.4:

TABLE 2.4. The log-linear weights for the case I=J=3.

$w_{ij,p}$	p=1	2	3	4
ij=11	1	0	1	0
12	1	0	0	1
13	1	0	0	0
21	0	1	1	0
22	0	1	0	1
23	0	1	0	0
31	0	0	1	0
32	0	0	0	1

From (2.44) we can then directly derive the likelihood equations as
$$\sum_i \sum_j w_{ij \cdot p} x_{ij} = \sum_i \sum_j w_{ij \cdot p}(n\pi_{ij}), \ p = 1,...,I+J-1.$$

Table 2.4 then shows that for p = 1, 2, 3 and 4 the likelihood equations are

$$x_{i \cdot} = n\pi_{i \cdot}, \ i = 1,2 \qquad (2.50)$$

and

$$x_{\cdot j} = n\pi_{\cdot j}, \ j = 1,2. \qquad (2.51)$$

From $\sum_i x_{i \cdot} = \sum_j x_{\cdot j} = n$ and $\sum_i \pi_{i \cdot} = \sum_j \pi_{\cdot j} = 1$ follows that (2.50) and (2.51) are also true for i=j=3. Hence the estimates for the marginal probabilities are

$$\hat\pi_{i \cdot} = \frac{x_i}{n}, \ i = 1,2,3 \qquad (2.52)$$

and

$$\hat\pi_{\cdot j} = \frac{x_{\cdot j}}{n}, \ j = 1,2,3. \qquad (2.53)$$

Under the hypothesis the expected numbers are accordingly the well-known expressions

BASIC THEORY

$$n\hat{\pi}_{ij} = \frac{X_{i\cdot}X_{\cdot j}}{n}.$$

The Z-test statistic is computed in the well-known way as

$$Z = 2 \sum \text{observed } (\ln(\text{observed}) - \ln(\text{expected})).$$

We thus get

$$Z(H_0) = 2 \sum_{i=1}^{I} \sum_{j=1}^{J} X_{ij} \left[\ln X_{ij} - \ln\left(\frac{X_{i\cdot}X_{\cdot j}}{n}\right) \right].$$

The asymptotic χ^2-distribution of $Z(H_0)$ has $(I-1)(J-1)$ degrees of freedom, since there are $IJ-1$ unconstrained multinomial probabilities in the saturated model and $I+J-2$ parameters to be estimated under the hypothesis, which means

$$df = IJ - I - J + 1 = (I-1) \cdot (J-1).$$

It is relatively easy to derive the matrix **M** needed to compute variances, standard errors and residuals.

In order to compute the inverse of **M** one really needs a computer, but the result is rather simple. The diagonal element (2.47) of the matrix **H** is thus (remembering that the elements in a two-way table have double subscripts)

$$h_{ij\cdot ij} = \frac{\pi_{i\cdot} + \pi_{\cdot j} - 2\pi_{ij}}{1 - \pi_{ij}},$$

where $\pi_{ij} = \pi_{i\cdot}\pi_{\cdot j}$. It follows that

$$\text{var}[\,X_{ij} - n\hat{\pi}_{ij}\,] = n\pi_{ij}(1 - \pi_{ij})(1 - h_{ij\cdot ij}) = n\pi_{ij}(1 - \pi_{i\cdot})(1 - \pi_{\cdot j}), \quad (2.54)$$

such that the standardised residuals under the independence hypothesis are given by

$$r_{ij} = \frac{X_{ij} - n\hat{\pi}_{ij}}{\sqrt{\text{var}[X_{ij} - n\hat{\pi}_{ij}]}}$$

with $\text{var}[\,X_{ij} - n\hat{\pi}_{ij}\,]$ given by (2.54).

Strictly speaking, we have not shown that the χ^2 approximation to the Z-test statistic is valid. To do so, we would have to verify that the assumptions for Theorem 2.5 are satisfied. Theorem 2.5 is the appropriate theorem to apply, since H_0 places constraints on the canonical parameters, as shown by (2.49). We must, therefore, show that there is reparametrization, which has the I+J-2 θ's as part of the new parameter set. This can be done although the expressions are not obvious and require some space to write. In addition it is not the most convenient parametrization if we want to generalize to multi-way contingency tables.

For these and other reasons, it is now standard for contingency tables to use a reparametrization which has the following form

$$\ln \pi_{ij} = \tau_{ij}^{AB} + \tau_i^A + \tau_j^B + \tau_0$$

with the linear constraints

$$\tau_{i\cdot}^{AB} = \sum_{j=1}^{J} \tau_{ij}^{AB} = 0 ,$$

$$\tau_{\cdot j}^{AB} = \sum_{i=1}^{I} \tau_{ij}^{AB} = 0 ,$$

$$\tau_{\cdot}^{A} = \sum_{i=1}^{I} \tau_i^A$$

and

$$\tau_{\cdot}^{B} = \sum_{j=1}^{J} \tau_j^B .$$

These constraints are necessary, because otherwise the model would be over-parametrized. Actually we have IJ-1 unconstrained π's and, due to the constraints, (I-1)(J-1) τ^{AB}'s, I-1 τ^A's and J-1 τ^B's, which adds up to IJ-1 parameters. Hence the last parameter τ_0 must be a function of the other parameters, which can also be seen from the expression for $\ln \pi_{ij}$ and the fact that $\Sigma \pi_{ij} = 1$. That we have a true reparametrization can be verified directly by calculating the τ's in terms of the $\ln \pi_{ij}$'s. We get for example

$$\tau_{ij}^{AB} = \lambda_{ij} - \bar{\lambda}_{i\cdot} - \bar{\lambda}_{\cdot j} + \bar{\lambda}_{\cdot\cdot} ,$$

where $\lambda_{ij} = \ln \pi_{ij}$, $\bar{\lambda}_{i\cdot} = \frac{1}{J}\sum_{j=1}^{J} \lambda_{ij}$, $\bar{\lambda}_{\cdot j} = \frac{1}{I}\sum_{i=1}^{I} \lambda_{ij}$ and $\bar{\lambda}_{\cdot\cdot} = \frac{1}{IJ}\sum_{i=1}^{I}\sum_{j=1}^{J} \lambda_{ij}$.

The most important property of this parametrization is that the hypothesis of independence is equivalent to

$$H_0 : \lambda_{ij}^{AB} = 0 \text{ for all } i \text{ and } j .$$

BASIC THEORY 35

In addition

$$\ln \pi_{ij} = \tau_i^A + \tau_j^B + \tau_0$$

under the hypothesis, as is easily seen.

2.11 The numerical solution of the likelihood equations for the log-linear model

The likelihood equations for a log-linear model are in almost all computer packages solved by the so-called **iterative proportional fitting method** or the **Deming-Stephan method**. This method usual requires more iterations than standard methods, like the Newton-Raphson method or the Fisher scoring method. But this is more than compensated for by the fact that the calculations, to be carried out in each iteration, are extremely simple.

Since the likelihood equations for the log-linear model according to (2.44) has the form

$$t_p^* = n \sum_j w_{jp} \pi_j \, , \, p = 1,...,m \, , \quad (2.55)$$

the vector of ML-estimates $\hat{\theta}$ for θ are found when $\pi_j(\hat{\theta}) = \hat{\pi}_j$ satisfies equations (2.55). The iterative proportional fitting method is, therefore, a method for obtaining values of the π's which satisfy (2.55). Sometimes this is all one need, for example to compute Z-test statistics. If the θ-estimates are required, the relationship (2.41) is used. On matrix form (2.41) can be written

$$\mathbf{B} = \mathbf{W}\theta \, , \quad (2.56)$$

where **B** has elements $\ln \pi_j - \ln \pi_k$.

The solution to (2.56) is

$$\theta = (\mathbf{W'W})^{-1}\mathbf{W'B} \, ,$$

from which the ML-estimates for the θ's are obtained, if the ML-estimates for the π's have already been calculated.

In order to obtain the π's which satisfies (2.55), the π's are adjusted p times in each iteration through the following adjustment formulas:

For p=1 we adjust $\pi_1,...,\pi_k$ proportionally as follows

$$\pi_{j1} = \frac{t_1^* \pi_{j0}}{n \sum_q w_{q1} \pi_{q0}} \, . \quad (2.57)$$

where $\pi_{11},\ldots,\pi_{k1}$ are the adjusted values in step 1 of iteration 1 and $\pi_{10},\ldots,\pi_{k0}$ are a set of initial values. In step p=2 we use almost the same formula, namely

$$\pi_{j2} = \frac{t_2^* \pi_{j1}}{n\sum_q w_{q2}\pi_{q1}} . \qquad (2.58)$$

to obtain new adjusted π-values $\pi_{12},\ldots,\pi_{k2}$.

In step p the adjustment formula is

$$\pi_{jp} = \frac{t_p^* \pi_{jp-1}}{n\sum_q w_{qp}\pi_{qp-1}} . \qquad (2.59)$$

It is clear, that the first likelihood Equation (2.55) for p=1 is satisfied if $\pi_{11},\ldots,\pi_{k1}$ satisfies (2.57). In the same way (2.55) is satisfied for p=2 if $\pi_{12},\ldots,\pi_{k2}$ satisfies (2.58). And in general (2.55) is satisfied for p=2 if $\pi_{1p},\ldots,\pi_{kp}$ satisfies (2.59). When steps 1 through m have been executed the last obtained adjustments $\pi_{1m},\ldots,\pi_{km}$ are compared to the initial values, and step 1 to m is repeated if the differences are above a preset limit, for example 0.0001. The procedure is now to continue repeating step 1 to m until the π's no longer change, for example if the changes are all numerically below 0.0001.

The iterative proportional fitting method is both extremely fast and has in addition the very useful feature that the choice of initial values are of no real consequence. Usually the trivial initial values $\pi_j = 1/m$ are chosen.

As mentioned the method produces a set of estimated multinomial probability parameters π_j, j=1,...m and therefore also a set of estimated expected cell counts $n\hat{\pi}_1,\ldots,n\hat{\pi}_k$, which satisfies the likelihood equations. Since only the $\hat{\pi}_j$'s are required to compute the value of a Z-test statistic the method directly produces the wanted result. If the ML-estimates for the θ's are required, the vector $\hat{\theta}$ of ML-estimates are obtained by inserting the $\hat{\pi}$'s in the solution to (2.56), that is

$$\hat{\theta} = (W'W)^{-1}W'\hat{B} ,$$

where the matrix $\hat{B}$ has elements $\ln\hat{\pi}_j - \ln\hat{\pi}_k$.

In the next chapter, we shall see how the method works for a 3-way contingency table.

2.12 Bibliographical notes

The literature on the exponential family goes far back. It has been attributed to many people and was for a long time known as the Fisher-Darmois-Pitman-Koopman model. As a basic model for making statistical inference it was first dis-

BASIC THEORY 37

cussed in depth by Lehmann (1959). The exact definition of the family and its key concepts as well as the main results in section 2.3 was established by Barndorff-Nielsen (1973) and published internationally in Barndorff-Nielsen (1978), see also Andersen (1980), Chapter 3. The asymptotic theory for test and estimators in exponential families was not given in Barndorff-Nielsen (1978), but is contained in Andersen (1980), building on Barndorff-Nielsens work, and unpublished results by A.H. Andersen.

The log-linear model was introduced by Birch (1973) as a model for three-way contingency tables. The connection between the multinomial distribution and the Poisson distribution in connection with log-linear models was first noted by Haberman (1974). The log-linear models, as defined in section 2.9, are special cases of a general class of models called the generalized linear model. It was introduced by Nelder and Wedderburn (1972). McCullagh and Nelder (1983) contains a systematic treatment of generalized linear models.

The iterative proportional fitting method was introduced by Deming and Stephan (1940).

2.13 Exercises

2.1 Let X follow the **geometric distribution** with point probability

$$f(x|\theta) = \theta^x(1-\theta) \ , \ x \geq 0 \ .$$

(a) Show that the geometric distribution belongs to the family of exponential distributions with canonical parameter $\tau = \ln(\theta)$ and $g(x) = x$.

(b) For n independent random variables $X_1, \ldots, X_n$ from a geometric distribution derive the ML-estimate for both τ and θ. [Hint: The mean value of X for a geometric distribution is $E[X] = \theta/(1-\theta)$.]

(c) Show that

$$K(\tau) = -\ln(1-e^\tau)$$

and use this result to derive the asymptotic variance of τ.

2.2 Consider again the geometric distribution in exercise 2.1. Show that the log-likelihood function is

$$\ln L(\tau) = \left(\sum_{i=1}^{n} x_i\right)\tau + n\ln(1-e^\tau).$$

(a) Use this result to derive the likelihood ratio test for $H_0 : \tau = \tau_0$.

(b) Let the observed values for n = 10 be 2, 4, 1, 0, 2, 1, 1, 2, 3, 1. Estimate τ and test the hypothesis $\tau = -0.3$.

2.3 Show by a direct calculation of the joint conditional probability that if $X_1,...,X_k$ are independent, Poisson distributed with parameters $\lambda_1,...,\lambda_k$ the distribution of $X_1,...,X_k$ given the sum $\Sigma X_i = n$ is multinomial with parameters n and

$$\pi_j = \frac{\lambda_j}{\lambda_.}, \ j=1,...,k .$$

2.4 Let the possible values of x be $j = 1,...,k$, and let $g_j(x) = 1$ if $x=j$ and $= 0$ if $x \neq j$. Then show that $P(X=x)$ can be written as

$$P(X=x) = \prod_{j=1}^{k} \pi_j^{g_j(x)} ,$$

where $\pi_j = P(X=j)$ and that

$$\ln P(X=x) = \sum_{j=1}^{k-1} g_j(x)\tau_j + \ln \pi_k ,$$

where $\tau_j = \ln \pi_j - \ln \pi_k$. Use this result to verify that the multinomial distribution belongs to the class of exponential families.

2.5 Let $X_1, ... ,X_n$ be independent Poisson distributed random variables with common mean λ.

(a) Show that the K-function is $K(\tau) = \exp(\tau)$.

(b) Use the K-function to derive the ML-estimate for τ from Equation (2.9).

(c) Derive $\text{var}[\hat{\tau}]$ from the derivatives of K.

An often used approximation when $\tau = g(\lambda)$ and g is a monotone function is

$$\text{var}[\hat{\tau}] \approx \left(g'(\lambda)\right)^2 \cdot \text{var}[\hat{\lambda}] ,$$

where g' is the derivative of g.

(d) In the Poisson distribution $\text{var}[X] = \lambda$. Verify that the approximation just mentioned, gives the same variance as the one you derived in (c).

2.6 In the Danish popular historical magazine Skalk, 1990, nr. 1, there is an article "The executioners axe" from which one can derive the number of executions for each decade in the 19th century.

Decades 1800-1900	1800 -1810	10 -20	20 -30	30 -40	40 -50	50 -60	60 -70	70 -80	80 -90	1890 -1900
Number of executions	10	46	18	19	17	19	5	0	2	1

(a) Argue that given the 137 executions in Denmark from 1800 to 1900 the numbers per decade can be described by a multinomial distribution of dimension k = 10 and probability parameters π_j, j=1,...,10.

(b) Test the hypothesis H_0 that the 137 executions are uniformly distributed over decades.

(c) Use the following information to formulate an alternative hypothesis H_1: (1) In 1866 a law was passed by the Danish Parliament, which greatly reduced the offenses for which the accused received a death penalty. But already in the early 60's the coming law was very much in the King's mind. (He was the only one who could grant clemency.) (2) In 1817 there was a major riot in the main Copenhagen prison. After the riot, 14 persons were given a death sentence. The King was very upset over the riot and refused to grant clemency for any of the convicted prisoners. Usually the King was rather lenient with granting clemency.

(d) Estimate the π's under the alternative hypothesis H_1 and test H_1.

2.7 The table shows the number of persons killed in the traffic in Denmark between 1981 and 1990.

	Year									
	81	82	83	84	85	86	87	88	89	90
Number killed	478	456	461	466	572	509	461	492	472	438

Test the hypothesis that the probability of being killed in the traffic has not changed over the decade 1981-1990 in Denmark.

2.8 Consider a trinomial distribution with probability parameters π_1, π_2 and π_3. We want to test the hypothesis

$$H_0 : \frac{\pi_1}{\pi_2} = \frac{\pi_2}{\pi_3} = 2 .$$

(a) Show that H_0 is equivalent to

$$\pi_1 = \frac{4}{7} , \quad \pi_2 = \frac{2}{7} , \quad \pi_3 = \frac{1}{7} ,$$

and test the hypothesis.

(b) Show that $\theta_1 = \tau_1 - \tau_2$, $\theta_2 = \tau_2$ is a reparametrization of the canonical parameters

$$\tau_1 = \ln \pi_1 - \ln \pi_3$$

and

$$\tau_2 = \ln \pi_2 - \ln \pi_3 .$$

Then show that H_0 in terms of the θ's can be expressed as

$$H_0 : \theta_1 = \theta_2 = \ln 2 .$$

2.9 Among 132 twins the distribution over the three combinations girl-girl (GG), girl-boy (GB) and boy-boy (BB) was

	GG	GB	BB	Total
Number	45	53	34	132

(a) If girls and boys are born independently with probability 0.5 for each possibility, the expected frequencies for the three combinations should be 1/4, 1/2 and 1/4. Test this hypothesis.

Let θ be the probability that a pair of twins being monozygotes (born from one egg rather than from two separate eggs), in which case both twins has the same sex.

(b) Show that the probability of the combinations GG and BB are both

$$P(BB) = P(GG) = \frac{1}{4}(1+\theta)$$

and

$$P(GB) = \frac{1}{2}(1-\theta) .$$

The ML-estimate for θ is

$$\hat{\theta} = \frac{x_{GG} + x_{BB} - x_{GB}}{n} .$$

(c) Estimate the percentages of monozygotes from the given data.

(d) Test that a model which allows for both monozygotes and dizygotes fits the data.

2.10 Show that for a 2×2 table the variance of the residuals for cells (1,1) and (2,2) under the independenced hypothesis are

BASIC THEORY

$$\text{var}[X_{11} - n\hat{\pi}_{11}] = n\pi_{11}\pi_{2.}\pi_{.2}$$

and

$$\text{var}[X_{22} - n\hat{\pi}_{22}] = n\pi_{22}\pi_{1.}\pi_{.1} .$$

[Hint: Use Equation (2.54).].

2.11 Consider a 2×2 table. Under the independence hypothesis we can introduce the parameters

$$\theta_1 = \ln\pi_{1.} - \ln\pi_{2.}$$

and

$$\theta_2 = \ln\pi_{.1} - \ln\pi_{.2}$$

Show that a log-linear model is defined by the log-linear weights $w_{ij\,p}$ in the table below

$w_{ij.p}$	p=1	2
ij=11	1	1
12	1	0
21	0	1

2.12 (a) Show that the formula

$$\tau_{ij}^{AB} = L_{ij} - \bar{L}_{i.} - \bar{L}_{.j} + \bar{L}_{..}$$

for a 2×2 table with $I = J = 2$ can be written as

$$\ln\pi_{11} - \frac{1}{2}(\ln\pi_{11} + \ln\pi_{12}) - \frac{1}{2}(\ln\pi_{11} + \ln\pi_{21}) + \frac{1}{4}\sum_{i=1}^{2}\sum_{j=1}^{2}\ln(\pi_{ij})$$

$$= \frac{1}{4}\Big[4\ln\pi_{11} - 2\ln(\pi_{11}\cdot\pi_{12}) - 2\ln(\pi_{11}\cdot\pi_{21}) + \ln(\pi_{11}\cdot\pi_{12}\cdot\pi_{21}\cdot\pi_{22})\Big]$$

(b) Use this result to show that for a 2×2 table the interaction τ_{11}^{AB} can be written

$$\tau_{11}^{AB} = \frac{1}{4}\ln\left(\frac{\pi_{11}\pi_{22}}{\pi_{12}\pi_{21}}\right).$$

Chapter 3

Three-way contingency tables

3.1 Log-linear models

A three-way contingency table is an array of observed values x_{ijk}, $i=1,...,I$, $j=1,...,J$, $k=1,...,K$ of I×J×K random variables, arranged in I rows, J columns and K layers. As model for the corresponding random variables, we choose

$$X_{111},...,X_{IJK} \sim M(n;\pi_{111},...,\pi_{IJK}) , \qquad (3.1)$$

that is a multinomial distribution with number parameter n and probability parameters π_{ijk}, where

$$n = x_{...} = \sum_i \sum_j \sum_k x_{ijk}$$

A three-way contingency table is often formed as one of the data summaries from a questionnaire given to a random sample of n individuals. In such a questionnaire the three-way contingency table is the cross-classification of the answers to three of the questions in the questionnaire, having, respectively, I, J and K response categories. If the sample is randomly drawn, and one respondent does not influence the responses of any other respondent, we can assume as statistical model, that the number of individuals x_{ijk} having response i on question 1, response j on question 2 and response k on question 3, follows the multinomial distribution (3.1), where the probability parameter π_{ijk} is the probability that a randomly drawn individual has his or her response in cell (ijk). If the sample is drawn from a well-defined population, this probability is also the frequency in the total population, who, if asked, would have responded in response categories i, j and k.

A terminology, which is convenient, is to introduce three **categorical variables** A, B and C. Variable A has I categories, variable B has J categories and variable C has K categories. If the response of one of the n individuals in the sample falls in cell (ijk), we shall say that variable A has observed value i, variable B observed value j and variable C observed value k.

EXAMPLE 3.1. *Children's use of video.*
In 1993 the Danish National Institute of Social Research investigated the way young people between 7 and 15 years old used their time outside school hours. Table 3.1 is from this investigation. The 3-way table shows how often the young people watch video, in three frequency categories, where the other categorical variables are sex and age.

TABLE 3.1. Frequency of watching videos at home or at friends homes for young people between 7 and 15 years of age, cross-classified according to age and sex.

Frequency of watching videos	Age	Sex	
		Boys	Girls
Almost daily	7 - 9 years	5	5
	10 - 12 years	4	3
	13 - 15 years	5	7
Every week	7 - 9 years	28	14
	10 - 12 years	20	17
	13 - 15 years	27	33
Seldom or never	7 - 9 years	88	87
	10 - 12 years	78	93
	13 - 15 years	100	81

Source: Andersen, D. (1995): School children's leisure hours. (In Danish). Report no. 95:2. Copenhagen: Danish National Institute of Social Research.

The three categorical variables forming the table are

A: *Frequency of watching videos.*

B: *Age.*

C: *Sex.*

The dimension of Table 3.1 is $I = 3$, $J = 3$ and $K = 2$.

As a help for grasping the formulas in the following, the Table 3.1 is shown as Table 3.2 with the general notations of this book.

TABLE 3.2. The notations of a general three-way contingency table of dimension 3×3×2.

Variable A	Variable B	Variable C	
	j=1	x_{111}	x_{112}
i=1	2	x_{121}	x_{122}
	3	x_{131}	x_{132}
	j=1	x_{211}	x_{212}
2	2	x_{221}	x_{222}
	3	x_{231}	x_{232}
	j=1	x_{311}	x_{312}
3	2	x_{321}	x_{322}
	3	x_{331}	x_{333}

We know from section 2.5 that the canonical parameters for the multinomial distribution (3.1) are $\tau_{ijk} = \ln(\pi_{ijk}) - \ln(\pi_{IJK})$. This parametrization has proved to be inferior to the log-linear parametrization, introduced briefly in section 2.10, for the purpose of testing the most important hypotheses for three-way contingency table. The **log-linear parametrization** for a 3-way contingency table is

$$\ln(n\pi_{ijk}) = \tau_0 + \tau_i^A + \tau_j^B + \tau_k^C + \tau_{ij}^{AB} + \tau_{ik}^{AC} + \tau_{jk}^{BC} + \tau_{ijk}^{ABC} . \qquad (3.2)$$

As $E[X_{ijk}] = n\pi_{ijk}$, the log-linear parametrization is a parametrization of the log-mean values in the cells. If we make the parameter change $\tau_0^* = \tau_0 - \ln(n)$, the corresponding parametrization for $\ln(\pi_{ijk})$ is

$$\ln(\pi_{ijk}) = \tau_0^* + \tau_i^A + \tau_j^B + \tau_k^C + \tau_{ij}^{AB} + \tau_{ik}^{AC} + \tau_{jk}^{BC} + \tau_{ijk}^{ABC} . \qquad (3.3)$$

The model has too many parameters - is over-parametrized - if no constraints are placed on the parameters, since 1+I+J+K+IJ+IK+JK+IJK > IJK. This over-parametrization can be overcome by introducing the linear constraints

$$\tau_{.}^A = \tau_{.}^B = \tau_{.}^C = 0 , \qquad (3.4)$$

$$\tau_{i.}^{AB} = \tau_{i.}^{AC} = \tau_{.j}^{AB} = \tau_{.k}^{AC} = \tau_{j.}^{BC} = \tau_{.k}^{BC} = 0 \qquad (3.5)$$

and

$$\tau^{ABC}_{ij\cdot} = \tau^{ABC}_{i\cdot k} = \tau^{ABC}_{\cdot jk} = 0 , \qquad (3.6)$$

where a dot ("·") means that the parameter has been summed over the index in question. A popular phrase is to say, that the log-linear parameters "sum to 0 over all indices". With these constraints, the parametrization (3.2) is a true reparametrization in the sense that the τ's can be derived in a unique way from the π's. For example

$$\tau^{AB}_{ij} = \bar{\mu}^*_{ij\cdot} - \bar{\mu}^*_{i\cdot\cdot} - \bar{\mu}^*_{\cdot j\cdot} + \bar{\mu}^*_{\cdot\cdot\cdot} ,$$

where $\mu^*_{ijk} = \ln(n\pi_{ijk})$, a dot again means a summation over the index in question, and a bar denotes an average. As an example of this notation

$$\bar{\mu}^*_{ij\cdot} = \frac{1}{K}\sum_{k=1}^{K} \mu^*_{ijk} .$$

All log-linear parameters have special names. Thus τ^{ABC}_{ijk} is a **three-factor interaction**, while τ^{AB}_{ij}, τ^{AC}_{ik} and τ^{BC}_{jk} are **two-factor interactions** and τ^{A}_{i}, τ^{B}_{j} and τ^{C}_{k} are **main effects**.

A consequence of the linear constraints (3.4) to (3.6) is that certain log-linear parameters are functions of other log-linear parameters. If, for example, τ^{ABC}_{ijk} is given for $i = 1,...,I-1$ and $j = 1,...,J-1$, the remaining values for $i = I$ and $j = J$ can be derived from (3.5) as

$$\tau^{AB}_{Ij} = -\sum_{i=1}^{I-1} \tau^{AB}_{ij}$$

$$\tau^{AB}_{iJ} = -\sum_{j=1}^{J-1} \tau^{AB}_{ij} ,$$

and

$$\tau^{AB}_{IJ} = -\sum_{i=1}^{I-1}\sum_{j=1}^{J-1} \tau^{AB}_{ij} .$$

It follows that only $(I-1)(J-1)$ 2-factor interactions between A and B have an unconstrained (or free) variation. Going through all log-linear parameters in this way, we get Table 3.3, which shows that the number of log-linear parameters with a free variation exactly equals the number of parameters in the multinomial distribution (3.1). Note, though, that τ_0 is a function of all other log-linear parameters due to the constraint

$$\sum_{i=1}^{I}\sum_{j=1}^{J}\sum_{k=1}^{K} \pi_{ijk} = 1 .$$

TABLE 3.3. Number of unconstrained (or free) log-linear parameters.

Parameter	Number
τ_i^A	$I - 1$
τ_j^B	$J - 1$
τ_k^C	$K - 1$
τ_{ij}^{AB}	$(I - 1)(J - 1)$
τ_{ik}^{AC}	$(I - 1)(K - 1)$
τ_{jk}^{BC}	$(J - 1)(K - 1)$
τ_{ijk}^{ABC}	$(I - 1)(J - 1)(K - 1)$
Total	$IJK - 1$

If all log-linear parameters are included in the model, we say that the model is **saturated**.

3.2 Log-linear hypotheses

Within the log-linear parametrization (3.2) a number of important hypotheses can be formulated by setting larger or smaller sets of τ's to 0. The following 7 are the **main types**.

H_1: $\tau_{ijk}^{ABC} = 0$ for all i, j and k
H_2: $\tau_{ijk}^{ABC} = \tau_{ij}^{AB} = 0$ for all i, j and k
H_3: $\tau_{ijk}^{ABC} = \tau_{ij}^{AB} = \tau_{ik}^{AC} = 0$ for all i, j and k
H_4^*: $\tau_{ijk}^{ABC} = \tau_{ij}^{AB} = \tau_{ik}^{AC} = \tau_i^A = 0$ for all i, j and k
H_4: $\tau_{ijk}^{ABC} = \tau_{ij}^{AB} = \tau_{ik}^{AC} = \tau_{jk}^{BC} = 0$ for all i, j and k
H_5: H_4 and $\tau_i^A = 0$ for all i
H_6: H_4 and $\tau_i^A = \tau_j^B = 0$ for all i and j
H_7: H_4 and $\tau_i^A = \tau_j^B = \tau_k^C = 0$ for all i, j and k

When we write "main types", it is because one - by exchange of letters - arrive at new hypotheses, but all such hypotheses are of the same type, in the sense that any hypothesis obtained from another by exchange of letters, can be treated statistically in exactly the same way. For this reason we study only the main types in the following discussion.

It is an important feature of log-linear models for 3-way tables, that all the main types of hypotheses, except H_1, can be interpreted as independence, conditional independence or uniform distribution over categories. In order to keep track of the various independencies and conditional independencies, we introduce the symbol

THREE-WAY CONTINGENCY TABLES

A ⊥ B for independence between A and B, the symbol A ⊥ B|C for conditional independence between A and B given C and A = u for a uniform distribution over the categories of A. Expressed in terms of the parameters of the multinomial distribution, these symbols correspond to the following.

(i) A and B are independent, and we write A ⊥ B, if and only if

$$\pi_{ij.} = \pi_{i..}\pi_{.j.} , \text{ for all i and j.} \qquad (3.7)$$

(ii) A and B are conditionally independent given C, and we write A ⊥ B|C, if and only if

$$\frac{\pi_{ijk}}{\pi_{..k}} = \frac{\pi_{i.k}}{\pi_{..k}} \frac{\pi_{.jk}}{\pi_{..k}}, \text{ , for all i, j and k,} \qquad (3.8)$$

that is if we have independence between variables A and B for every level k of variable C.

(iii) There is uniform distribution over the categories of A, and we write A = u, if and only if

$$\pi_{i..} = 1/I , \text{ for all i.} \qquad (3.9)$$

It is not particularly hard, although it requires a few lines of algebra, to show, that H_2 is satisfied if and only if (3.8) is satisfied. In the same way (but we again omit details) it can be shown that H_3 holds if and only if (3.7) is satisfied and at the same time

$$\pi_{i.k} = \pi_{i..}\pi_{..k} \text{ for all i and k .}$$

If one inspects all the other main types of hypotheses in the same manner, the following results emerge.

THEOREM 3.1. *Hypotheses H_2 through H_7 can all be interpreted as independence, conditional independence or uniform distribution over categories as follows*

H_2: A ⊥ B|C
H_3: A ⊥ B,C
H_4^*: A ⊥ B,C and A = u
H_4: A ⊥ B ⊥ C
H_5: A ⊥ B ⊥ C and A = u
H_6: A ⊥ B ⊥ C and A = B = u
H_7: A ⊥ B ⊥ C and A = B = C = u .

H_1 is a special case, because it cannot be interpreted in terms of independence, conditional independence or uniform distribution over categories.

With the interpretations in Theorem 3.1, we can write down explicit expressions for the cell probabilities under each of the hypotheses except H_1, as shown in Theorem 3.2.

THEOREM 3.2. *Under the hypotheses H_2 to H_7 the cell probabilities have the explicit expressions*

H_2: $\pi_{ijk} = \pi_{i.k}\pi_{.jk} / \pi_{..k}$
H_3: $\pi_{ijk} = \pi_{i..}\pi_{.jk}$
H_4^*: $\pi_{ijk} = \pi_{.jk} / I$
H_4: $\pi_{ijk} = \pi_{i..}\pi_{.j.}\pi_{..k}$
H_5: $\pi_{ijk} = \pi_{.j.}\pi_{..k} /I$
H_6: $\pi_{ijk} = \pi_{..k} / (IJ)$
H_7: $\pi_{ijk} = 1 / (IJK)$.

The relationships in Theorem 3.2 are easily derived successively from (3.7) to (3.9). The expression for H_2, for example, follows directly from (3.8). If we then rewrite $\pi_{i.k}$ according to (3.7) by changing the letters j and k and insert the resulting expression $\pi_{i.k} = \pi_{i..}\pi_{..k}$ in (3.8), we get H_3, etc. For H_1 there is no explicit expression for π_{ijk} as a function of the π-marginals.

In the next chapter we discuss in more detail the connection between the various hypotheses, the formulation of the various hypotheses in terms of log-linear parameters, and the interpretation of the hypotheses.

An **association diagram** for a log-linear model is a graph, where each variable is indicated by a point and the letter allocated to the variable. Two points in the associationdiagram are connected by a line if, under the given hypothesis, there is an interaction involving those two variables, which is not assumed to be 0. For H_1 all points in the diagram must, according to this rule, be connected, because all 3 two-factor interactions are included. For H_2, on the other hand, only A to C, and B to C need to be connected since all interactions involving both A and B are zero.
Figure 3.1 shows the association diagrams for all the eight hypotheses. A uniform distribution over categories is denoted by a star ("*").

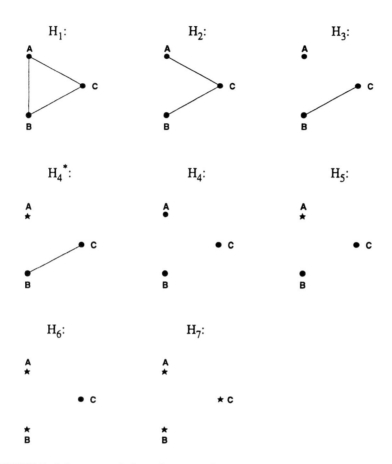

FIGURE 3.1. Association diagrams for hypotheses H_2 to H_7 in a three-way contingency table.

A more complete discussion of association diagrams is postponed to chapter 4. For the present, it suffices to say that two variables are independent, if they are not in any way connected in the association diagram, and that two variables are conditionally independent given a third variable, if they are unconnected directly in the association diagram, but connected by a route passing through this third variable, as shown for H_2 in Figure 3.1.

3.3 Estimation

Because the x's follow a multinomial distribution with probability parameters, whose logarithm is given by (3.2), the log-likelihood function is given by

$$\ln L = \text{const.} + \sum_i \sum_j \sum_k x_{ijk} \ln \pi_{ijk}$$

$$= \text{const.} + \tau_0 x_{...} + \sum_i \tau_i^A x_{i..} + \sum_j \tau_j^B x_{.j.} + \sum_k \tau_k^C x_{..k}$$

$$+ \sum_i \sum_j \tau_{ij}^{AB} x_{ij.} + \sum_i \sum_k \tau_{ik}^{AC} x_{i.k} + \sum_j \sum_k \tau_{jk}^{BC} x_{.jk} + \sum_i \sum_j \sum_k \tau_{ijk}^{ABC} x_{ijk},$$

where const. is a constant which is independent of the τ's. The model thus belongs to an exponential family, where the log-linear parameters are the canonical parameters and the sufficient statistics are different marginals derived from the contingency table. It then follows from Theorem 2.1 that the ML-estimates for the log-linear parameters are obtained by equating these marginals with their mean values. A complete estimation of all log-linear parameters in the saturated model is thus obtained by solving the following system of likelihood equations:

$$x_{...} = E[X_{...}] = n\pi_{...} = n, \qquad (3.10)$$

$$x_{i..} = E[X_{i..}] = n\pi_{i..}, \; i=1,...,I \qquad (3.11)$$

$$x_{.j.} = E[X_{.j.}] = n\pi_{.j.}, \; j=1,...,J \qquad (3.12)$$

$$x_{..k} = E[X_{..k}] = n\pi_{..k}, \; k=1,...,K \qquad (3.13)$$

$$x_{ij.} = E[X_{ij.}] = n\pi_{ij.}, \; i=1,...,I, \; j=1,...,J \qquad (3.14)$$

$$x_{i.k} = E[X_{i.k}] = n\pi_{i.k}, \; i=1,...,I, \; k=1,...,K \qquad (3.15)$$

$$x_{.jk} = E[X_{.jk}] = n\pi_{.jk}, \; j=1,...,J, \; k=1,...,K \qquad (3.16)$$

$$x_{ijk} = E[X_{ijk}] = n\pi_{ijk}, \; i=1,...,I, \; j=1,...,J, \; k=1,...,K. \qquad (3.17)$$

All these equations are satisfied if (3.17) is satisfied, that is if

$$\hat{\pi}_{ijk} = x_{ijk}/n, \qquad (3.18)$$

which is the trivial estimation of π_{ijk} in the saturated model.

Actually the equation system (3.10) to (3.17) is overdetermined in the sense that there are more equations than log-linear parameters, when we let all indices run to their upper limit. As an example (3.14) consists of I×J equations, while there are only $(I-1)(J-1)$ unconstrained τ_{ij}^{AB}'s. It is on the other hand, as we shall see below, inconvenient to restrict all likelihood equations to those with indices $i \leq I-1, j \leq J-1$ and $k \leq K-1$.

THREE-WAY CONTINGENCY TABLES 51

If we want to estimate the parameters under one of the hypotheses, only some of the Equations (3.10) to (3.17) must be solved. It is thus one of the important tasks, when estimating parameters, to keep track of which equations are included in the estimation process.

As mentioned, all equations are satisfied if π_{ijk} is estimated as (3.18). This means that under the saturated model the ML-estimates for the log-linear parameters are obtained by using (3.2), defining the τ's with τ_{ijk} replaced by $\hat{\pi}_{ijk} = x_{ijk}/n$, and then using the constraints (3.4) to (3.6).

If, for example, under the saturated model we need the ML-estimate for τ_{ij}^{AB}, we find from

$$L_{ijk} = \ln\left(\frac{x_{ijk}}{n}\right) = \ln(\hat{\pi}_{ijk}) ,$$

using (3.4) to (3.6) that

$$\bar{L}_{ij.} = \hat{\tau}_0^* + \hat{\tau}_i^A + \hat{\tau}_j^B + \hat{\tau}_{ij}^{AB} ,$$

$$\bar{L}_{i..} = \hat{\tau}_0^* + \hat{\tau}_i^A ,$$

$$\bar{L}_{.j.} = \hat{\tau}_0^* + \hat{\tau}_j^B ,$$

and

$$\bar{L}_{...} = \hat{\tau}_0^* ,$$

such that

$$\hat{\tau}_{ij}^{AB} = \bar{L}_{ij.} - \bar{L}_{i..} - \bar{L}_{.j.} + \bar{L}_{...} , \qquad (3.19)$$

where as before a bar means an average and a dot a summation over the index in question.

It is a little more complicated when the model is not saturated. Because all hypotheses H_1 to H_7 have the form that a certain subset of log-linear parameters are zero, ML-estimation of parameters under any of these hypotheses is the same as solving a subset of the likelihood Equations (3.10) to (3.17). It is also clear which equations to solve, namely those corresponding to log-linear parameters not assumed to be zero under the hypothesis. For the first hypothesis H_1, we must according to this rule solve Equations (3.10) to (3.16), but not (3.17) since τ_{ijk}^{ABC} is assumed to be zero. As another example, if we need to find the ML-estimates under H_3, Equations (3.10) to (3.13) and (3.16) must be solved, but not (3.14), (3.15) and (3.17). Matters are simplified by the fact that if certain equations are zero, several other equations are also satisfied automatically, as we saw with Equation (3.17), which entailed that all other Equations were satisfied. As a typical example, if (3.16) is satisfied then automatically (3.10), (3.12) and (3.13) are also satisfied. Below is a listing of those equations we need to solve for each hypothesis.

Hypothesis	Likelihood equations
H_1	(3.14), (3.15), (3.16)
H_2	(3.15), (3.16)
H_3	(3.16), (3.11)
H_4^*	(3.16)
H_4	(3.11), (3.12), (3.13)
H_5	(3.12), (3.13)
H_6	(3.13)
H_7	(3.10)

Note in particular that we need to add Equation (3.11) under H_3, since it does not follow from (3.16) that (3.11) is satisfied. In most cases we can omit equations which correspond to marginals with fewer indices than those marginals already included, but there are exceptions. It is even simple to describe when this happens; namely when a marginal of lower order has indices which are not a subset of the indices of the included marginals of higher order. This is what happened for hypothesis H_3. Here the only 2-factor marginal included in the likelihood equation is $x_{.jk}$, but the index "i" is missing, which is why we have to include the one factor marginal $x_{i..}$.

The likelihood equations to be solved can be expressed in symbolic form by writing down the combinations of variable names which corresponds to the marginals. Thus for H_1, we have to solve (3.14), (3.15) and (3.16) corresponding to the marginals $x_{ij.}$, $x_{i.k}$ and $x_{.jk}$. We write this as AB, AC, BC.

The marginals, which are equated to their mean values for a given hypothesis, are called **sufficient marginals**. For the hypothesis H_1 the sufficient marginals are, as we saw, $x_{ij.}$, $x_{i.k}$ and $x_{.jk}$. For the hypothesis H_3 the sufficient marginals are $x_{.jk}$ and $x_{i..}$, etc. In symbolic form the sufficient marginals for H_3 are BC and A. Table 3.4 gives a survey of the typical hypotheses, the sufficient marginals, their symbolic forms and the interpretation of the hypotheses.

THREE-WAY CONTINGENCY TABLES 53

TABLE 3.4. Sufficient marginals, symbolic forms and interpretations for hypotheses H_1 through H_7.

Hypothesis	Sufficient marginals	Symbolic form	Interpretation
H_1	$x_{ij.}\ x_{i.k}\ x_{.jk}$	AB, AC, BC	-
H_2	$x_{i.k}\ x_{.jk}$	AC, BC	$A \perp B \mid C$
H_3	$x_{.jk}\ x_{i..}$	BC, A	$A \perp B, C$
H_4^*	$x_{.jk}$	BC	$A \perp B, C$ and $A = u$
H_4	$x_{i..}\ x_{.j.}\ x_{..k}$	A, B, C	$A \perp B \perp C$
H_5	$x_{.j.}\ x_{..k}$	B, C	$A \perp B \perp C$ and $A = u$
H_6	$x_{..k}$	C	$A \perp B \perp C$ and $A = B = u$
H_7	$x_{...}$	-	$A = B = C = u$

Note that there is no interpretation for H_1.

Given the sufficient marginals for a hypothesis, the corresponding likelihood equations are solved with respect to the π's by the **iterative proportional fitting method** we introduced in section 2.11. This method is so fast and reliable that it is used even in cases where there are explicit expressions for the solutions. For 3-way tables this is the case for all hypotheses except H_1.

We now illustrate the iterative proportional fitting method for the solution of Equations (3.14) to (3.16), that is for hypothesis H_1. The method, we recall, result in estimates $\hat{\pi}_{ijk}$ for the cell probabilities under H_1. From (2.57) follows that the π's are adjusted in the first step from initial values $\pi_{ijk}^{(0)}$ by

$$\pi_{ijk}^{(1)} = \frac{1}{n} \frac{x_{ij.}\cdot\pi_{ijk}^{(0)}}{\pi_{ij.}^{(0)}} . \qquad (3.20)$$

In this step we ensure that Equation (3.14) is satisfied, since (3.20) implies

$$n\pi_{ij.}^{(1)} = x_{ij.} .$$

From (2.58) follows that the $\pi_{ijk}^{(1)}$'s in step 2 are adjusted by

$$\pi_{ijk}^{(2)} = \frac{1}{n} \frac{x_{i.k}\pi_{ijk}^{(1)}}{\pi_{i.k}^{(1)}} . \qquad (3.21)$$

Then (3.15) is satisfied since now

$$n\pi_{i.k}^{(2)} = x_{i.k} .$$

In the third and last step of iteration 1 $\pi_{ijk}^{(2)}$ is adjusted by

$$\pi_{ijk}^{(3)} = \frac{1}{n} \frac{x_{.jk} \pi_{ijk}^{(2)}}{\pi_{.jk}^{(2)}}, \qquad (3.22)$$

to ensure that ensure that Equation (3.16) is satisfied.

Steps 1 through 3 are now repeated until the π's no longer change their values within a specified level, for example 0.0001. The resulting values of π_{ijk} are the ML-estimates $\hat{\pi}_{ijk}$, since now Equations (3.14), (3.15) and (3.16) are all satisfied.

The method usually converges very fast. In addition the method is almost completely insensitive to the choice of initial values. It is customary, therefore, to choose the initial values

$$\pi_{ijk}^{(0)} = \frac{1}{IJK}.$$

Alternative to stopping the iterations when the π's do not change any more within a specified limit, one can choose to stop when the expected values under the hypothesis do not change any more. Here a wise choice is a change less than 0.001, which would ensure that the expected values are accurate up to the second decimal point.

The conditions for a unique set of solutions to the likelihood equations are very simple for the log-linear model (3.2). The likelihood equations simply have a set of unique solutions if under a given hypothesis none of the sufficient marginals have the value zero. If there are zero's in one or more cells of a contingency table, we talk about **incomplete tables**. As there are only problems with the solution of likelihood equations when the sufficient marginals under a given hypothesis are zero, there may well be unique solutions also for incomplete tables and models not equivalent to the saturated model. Chapter 5 is devoted to a more thorough discussion of incomplete contingency tables

EXAMPLE 3.1 (continued). *We now demonstrate how the iterative proportional fitting procedure works for the data in Table 3.1 and ML-estimation under H_1. Table 3.5 is Table 3.1 extended to include all the marginals we need for solving likelihood equations.*

		Sex		
Frequency of watching videos	Age in years	Boys	Girls	Total
Almost daily	7 - 9	5	5	10
	10 - 12	4	3	7
	13 - 15	5	7	12
	Total	14	15	29
Every week	7 - 9	28	14	42
	10 - 12	20	17	37
	13 - 15	27	33	60
	Total	75	64	139
Seldom or never	7 - 9	88	87	175
	10 - 12	78	93	171
	13 - 15	100	81	181
	Total	266	261	527
Total	7 - 9	121	106	227
	10 - 12	102	113	215
	13 - 15	132	121	253
	Total	355	340	695

We start by setting all expected values equal to 695/18 = 38.61. Following (3.20) in step 1 we multiply $x_{ij.}$ by 38.61/(38.61+38.61) = 0.5, since k has two levels. This gives the expected numbers in the next table.

		Sex		
Frequency of watching videos	Age in years	Boys	Girls	Total
	7 - 9	5.0	5.0	10
Almost daily	10 - 12	3.5	3.5	7
	13 - 15	6.0	6.0	12
	Total	14.5	14.5	29
	7 - 9	21.0	21.0	42
Every week	10 - 12	18.5	18.5	37
	13 - 15	30.0	30.0	60
	Total	69.5	69.5	139
	7 - 9	87.5	87.5	175
Seldom or never	10 - 12	85.5	85.5	171
	13 - 15	90.5	90.5	181
	Total	263.5	263.5	527
	7 - 9	113.5	113.5	227
Total	10 - 12	107.5	107.5	215
	13 - 15	126.5	126.5	253
	Total	347.5	347.5	695

In step 2 the expected numbers in this table are, according to (3.21), adjusted by the factor

$$\frac{x_{i.k}}{\pi_{i.k}^{(1)}}$$

The number in cell (111) is thus multiplied by $14/14.5 = 0.9655$, giving 4.83. The expected numbers after two steps are

Three-way contingency tables

Frequency of watching videos	Age in years	Sex Boys	Girls	Total
	7 - 9	4.83	5.17	10.00
Almost daily	10 - 12	3.38	3.62	7.00
	13 - 15	5.79	6.21	12.00
	Total	14.00	15.00	29.00
	7 - 9	22.66	19.34	42.00
Every week	10 - 12	19.96	17.04	37.00
	13 - 15	32.38	27.62	60.00
	Total	75.00	64.00	139.00
	7 - 9	88.33	86.67	175.00
Seldom or never	10 - 12	86.31	84.69	171.00
	13 - 15	91.36	89.64	181.00
	Total	266.00	261.00	527.00
	7 - 9	115.82	111.18	227.00
Total	10 - 12	109.65	105.35	215.00
	13 - 15	129.53	123.47	253.00
	Total	355.00	340.00	695.00

Finally in step 3 the numbers in this table are, according to (3.22), adjusted with the factor

$$\frac{x_{.jk}}{\pi^{(2)}_{.jk}}$$

The number in cell (111) is thus multiplied by 121/115.82 = 1.0447, giving 5.05 The expected numbers after step three are shown in the table below together with the expected values after the final iteration, which in this case, using the BMDP package, was iteration number 3.

Sex

Frequency of watching videos	Age in years	Boys Step 3	Boys Final values	Girls Step 3	Girls Final values	Total Step 3	Total Final values
Almost daily	7 - 9	5.05	5.02	4.93	4.98	9.98	10.00
	10 - 12	3.14	3.11	3.88	3.89	7.02	7.00
	13 - 15	5.90	5.87	6.09	6.13	11.99	12.00
	Total	14.09	14.00	14.90	15.00	28.99	29.00
Every week	7 - 9	23.67	23.55	18.44	18.45	42.11	42.00
	10 - 12	18.57	18.59	18.27	18.41	36.84	37.00
	13 - 15	33.00	32.87	27.07	27.13	60.07	60.00
	Total	75.24	75.00	63.78	64.00	139.02	139.00
Seldom	7 - 9	92.28	92.43	82.63	82.57	174.91	175.00
	10 - 12	80.28	80.31	90.84	90.69	171.12	171.00
	13 - 15	93.10	93.26	87.85	87.74	180.95	181.00
	Total	265.66	266.00	261.32	261.00	526.98	527.00
Total	7 - 9	121.00	121.00	106.00	106.00	227.00	227.00
	10 - 12	101.99	102.00	112.99	113.00	214.98	215.00
	13 - 15	132.00	132.00	121.01	121.00	253.01	253.00
	Total	354.99	355.00	340.00	340.00	694.99	695.00

As is easily seen, the first cycle of three steps has already brought us close to the solutions, in spite of the fact that we started out with the same expected number in each cell. Thus only a few more iterations of step 1 to step 3 are required, in this case 2.

Although the iterative proportional fitting procedure is almost always used, it is for 3-way tables only necessary for the estimation of parameters under H_1. For all other models there is an explicit solution to the likelihood equations. These solutions follow from Theorem 3.2 when we replace the π's on the right hand sides with the corresponding sufficient marginals divided by the sample size n, for example $\pi_{i.k}$ with $x_{i.k}/n$, $\pi_{i..}$ with $x_{i..}/n$ etc. Then the π's on the left hand side obviously satisfy the likelihood equations, since the sufficient marginals for a given hypothesis is equal to its mean value. As an example (3.15) and (3.16), which are the likelihood equations H_2, are satisfied for

$$\hat{\pi}_{ijk} = (x_{i.k} x_{.jk}) / (n x_{..k}),$$

since Equation (3.15) becomes

THREE-WAY CONTINGENCY TABLES

$$x_{i.k} = n(x_{i.k} x_{..k}) / (nx_{..k})$$

and Equation (3.16) becomes

$$x_{.jk} = n(x_{..k} x_{.jk}) / (nx_{..k}).$$

The validity of the remaining relationships can be verified in the same way. A survey of these results is given in Table 3.5.

TABLE 3.5. Estimated cell probabilities under hypotheses H_2 through H_7.

Hypothesis	Estimated cell probability
H_2	$\hat{\pi}_{ijk} = (x_{i.k} x_{.jk}) / (nx_{..k})$
H_3	$\hat{\pi}_{ijk} = (x_{i..} x_{.jk}) / n^2$
H_4^*	$\hat{\pi}_{ijk} = x_{.jk} / (nI)$
H_4	$\hat{\pi}_{ijk} = (x_{i..} x_{.j.} x_{..k}) / n^3$
H_5	$\hat{\pi}_{ijk} = (x_{.j.} x_{..k}) / (n^2 I)$
H_6	$\hat{\pi}_{ijk} = x_{..k} / (nIJ)$
H_7	$\hat{\pi}_{ijk} = 1 / (IJK)$

3.4 Testing hypotheses

For a 3-way contingency table, there are seven prototypes of hypotheses we may want to test, namely H_1 through H_7. In order to do so, we apply Theorem 2.4, since all hypotheses can be formulated in terms of canonical parameters being set equal to 0 within a log-linear model. Equation (2.30) then shows that all we have to do is calculate the value of $-2 \cdot \ln(L)$ with the canonical parameters, estimated under the hypothesis, inserted and with the canonical parameters, estimated without the hypothesis, inserted. This task is greatly facilitated by the fact, that $-2 \ln(L)$ for the multinomial distribution (3.1) is

$$-2 \ln L = -2 \sum_i \sum_j \sum_k x_{ijk} \ln \pi_{ijk}.$$

Accordingly it suffices to estimate the cell probabilities under and without the hypothesis. For the purpose of testing hypotheses we need not bother with deriving the estimates for the log-linear parameters, that is the τ's, which as we have seen are functions of the estimated cell probabilities. It follows that if $\hat{\pi}_{ijk}$ are the estimated cell probabilities under the hypothesis H, and $\hat{\pi}_{ijk}$ the estimated cell probabilities without this hypothesis, the test statistic (2.30) takes the form

$$Z(H) = 2\sum_i \sum_j \sum_k X_{ijk}(\ln \hat{\pi}_{ijk} - \ln \tilde{\pi}_{ijk}) . \qquad (3.23)$$

Z(H) is under H, according to Theorem 2.4, approximately distributed as a χ^2-distributed random variable with degrees of freedom equal to the number of unconstrained log-linear parameters which are set to zero under H. Without H, that is in the saturated model, the ML-estimates for the cell probabilities are $\hat{\pi}_{ijk} = x_{ijk}/n$ so that (3.23) takes the form

$$Z(H) = 2\sum_i \sum_j \sum_k X_{ijk}\left(\ln X_{ijk} - \ln(n\tilde{\pi}_{ijk})\right) . \qquad (3.24)$$

The test statistics thus have the well-known form

$$2\sum_i \sum_j \sum_k \text{observed} \cdot \left(\ln(\text{observed}) - \ln(\text{expected})\right) .$$

Note that it is not necessary to calculate the ML-estimates for the log-linear parameters if we only want to test hypotheses, since the Z-test statistic only depends on the estimated π's. In view of the iterated proportional fitting method, which provides estimates for the π's but not for the τ's, this is an obvious advantage.

The degrees of freedom for the approximating χ^2-distribution are the number of log-linear parameters specified to be 0. Hence it is easy, by simple counting, to find the number of degrees of freedom for testing each of the hypotheses H_1 to H_7. For H_1 there are (I-1)(J-1)(K-1) 3-factor interactions, which are assumed 0, which is then the number of degrees of freedom. For H_2, in addition, (I-1)(J-1) 2-factor interactions are 0, such that the number of degrees of freedom become

$$(I-1) \cdot (J-1) \cdot (K-1) + (I-1) \cdot (J-1) = (I-1) \cdot (J-1) \cdot K$$

All the degrees of freedom df(H) for the various hypotheses are listed in Table 3.6.

TABLE 3.6. Number of degrees of freedom df(H) for Z(H), H = $H_1,...,H_7$.

H	Parameters set equal to 0	df(H)
H_1	$\tau^{ABC}_{ijk} = 0$	$(I-1)(J-1)(K-1)$
H_2	$\tau^{ABC}_{ijk} = \tau^{AB}_{ij} = 0$	$(I-1)(J-1)K$
H_3	$\tau^{ABC}_{ijk} = \tau^{AB}_{ij} = \tau^{AC}_{ik} = 0$	$(I-1)(KJ-1)$
H_4^*	$\tau^{ABC}_{ijk} = \tau^{AB}_{ij} = \tau^{AC}_{ik} = \tau^A_i = 0$	$KJ(I-1)$
H_4	$\tau^{ABC}_{ijk} = \tau^{AB}_{ij} = \tau^{AC}_{ik} = \tau^{BC}_{jk} = 0$	$IJK - I - J - K + 2$
H_5	H_4 and $\tau^A_i = 0$	$IKJ - J - K + 1$
H_6	H_4 and $\tau^A_i = \tau^B_j = 0$	$IKJ - K$
H_7	H_4 and $\tau^A_i = \tau^B_j = \tau^C_k = 0$	$IKJ - 1$

In order to test the hypotheses we thus compute the test statistics

$$z(H) = 2\sum_i \sum_j \sum_k x_{ij}\left(\ln x_{ijk} - \ln(n\hat{\pi}_{ijk})\right), \quad (3.25)$$

where $n\hat{\pi}_{ijk}$ are the estimated expected numbers H, and reject H if

$$Z(H) > \chi^2_{1-\alpha}(df(H))$$

where α is the chosen level for the test, for example 0.05. If H is one of the types presented in the start of section 3.2, the number of degrees of freedom for Z is found in Table 3.6. If H is not one of the hypothesis in Table 3.6, the degrees of freedom are obtained by interchanging I, J and K in the same way as the letters A, B and C are interchanged.

It may happen that we only want to test one specific hypothesis; in which case the test statistic (3.25) and the test procedure just outlined can be used. But a far more common situation is that we want - among many candidates - to find a hypothesis, which can be accepted at a reasonable low test level, and for which the corresponding model has a simple and relevant interpretation. In order to search for such a hypothesis and its corresponding model, there are several possible strategies, which we shall study in more detail in section 3.6. Some important strategies are based on a sequential procedure, where the relative merits of two hypotheses H_0 and H_0^* are compared by means of the sequential test statistic (2.34). Before we apply (2.34), note that H_0 must be included in H_0^* in the sense that all τ's, which are 0 under H_0^*, also are 0 under H_0. To test H_0 against H_0^* is thus equivalent to testing the hypothesis that those τ's, which are not already assume 0 under H_0^*, are 0; that is we test that

$$\tau_j = 0, j = 1,...,r, \quad (3.26)$$

but not that $\tau_j = 0$, j =r+1,...,m. A sequential testing of H_0 against H_0^* is thus the same as testing (3.26) assuming that

$$\tau_j = 0 , j = r+1,\ldots,m . \tag{3.27}$$

We shall use the symbol $H \subset H^*$, for situations where all log-linear parameters, which are assumed 0 in H_0^* are also assumed 0 under H_0.

From the definitions given at the start of section 3.2 it is now clear that we have the relationships

and
$$H_7 \subset H_6 \subset H_5 \subset H_4 \subset H_3 \subset H_2 \subset H_1$$
$$H_7 \subset H_6 \subset H_5 \subset H_4^* \subset H_3 \subset H_2 \subset H_1 .$$

For any two hypotheses H and H^*, which according to these relationships satisfy $H \subset H^*$ it then follows from (2.34) and (3.25) that

$$Z(H|H^*) = Z(H) - Z(H^*)$$
$$= 2\sum_i \sum_j \sum_k X_{ijk}\left(\ln(n\hat{\pi}_{ijk}^*) - \ln(n\hat{\pi}_{ijk})\right), \tag{3.28}$$

where $\hat{\pi}_{ijk}$ are the estimated cell probabilities under H and $\hat{\pi}^*_{ijk}$ the estimated cell probabilities under H^*.

According to Theorem 2.6 $Z(H|H^*)$ is approximately χ^2-distributed with degrees of freedom

$$df(H|H^*) = df(H) - df(H^*) .$$

If we want to evaluate the goodness of fit of the hypothesis

$$H: \tau_{ik}^{AC} = 0 \quad \text{for all } i \text{ and } k ,$$

given that $\tau_{ijk}^{ABC} = \tau_{ij}^{AB} = 0$ for all i, j and k, we are testing H_3 against H_2, and the sequential test statistic to use is

$$Z(H_3|H_2) = 2\sum_i \sum_j \sum_k X_{ijk}\left(\ln(n\hat{\pi}_{ijk}^*) - \ln(n\hat{\pi}_{ijk})\right),$$

where according to Table 3.5

and
$$\hat{\pi}_{ijk}^* = (X_{i.k} X_{.jk}) / (nX_{..k})$$
$$\hat{\pi}_{ijk} = (X_{i..} X_{.jk}) / n^2 .$$

The number of degrees of freedom for $Z(H_3|H_2)$ is

$$df(H_3) - df(H_2) = (I - 1)(KJ - 1) - (I - 1)(J - 1)K = (I - 1)(K - 1) ,$$

which is also the number of unconstrained τ_{ik}^{AC}'s.

EXAMPLE 3.2. *Possesion of a freezer.*
One of the questions included in the 1976 Danish Welfare Study was whether there was a freezer in the household or not. Table 3.7 shows the cross-classification of this categorical variable, called A, with categories "yes" and "no", and variables B: Social group, with four categories (cf. Example 1.1) and C: Ownership, with categories "Owner" and "Renter", dependent on whether the interviewed owned his or her dwelling or rented it.

TABLE 3.7. The sample in the Danish Welfare Study cross-classified according to Possession of a freezer, Social group and Ownership

		C: Ownership	
A: Freezer	B: Social group	Owner	Renter
Yes	I-II	304	92
	III	666	174
	IV	894	379
	V	720	433
No	I-II	38	64
	III	85	113
	IV	93	321
	V	84	297

Source: Hansen,E.J. (1978): The Distribution of Living Conditions. Publication 82. Danish National Institute for Social Research. Copenhagen: Teknisk Forlag. The table is constructed from the data base available at the Danish Data Archives. Odense. Denmark.

Table 3.8 shows the test statistics for the hypotheses H_1 to H_7.

TABLE 3.8. Test statistics and levels of significance for H_1 to H_7 for the data in Table 3.7.

Hypothesis	Sufficient marginals	Z(H)	Degrees of freedom	Level of Significance
H_1	AB,AC,BC	5.97	3	0.113
H_2	AC,BC	7.38	6	0.287
H_3	BC,A	661.06	7	0.000
H_4	BC	2122.83	8	0.000
H_4	A,B,C	783.30	10	0.000
H_5	B,C	2245.08	11	0.000
H_6	C	3056.32	14	0.000
H_7	-	3094.51	15	0.000

From Table 3.8 it follows that hypotheses H_1 and H_2 can be accepted, while all the remaining hypotheses are rejected. H_2 has the association diagram shown in Figure 3.2.

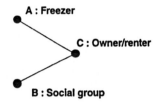

FIGURE 3.2. Association diagram for H_2.

The interpretation of H_2 is $A \perp B | C$; that is there is no association between position according to social group and possession of a freezer, if we look at owners and renters separately.

According to Table 3.5 we have explicit expressions for the estimated cell probabilities under H_2. The expected numbers under H_2 are thus easily derived. The expected number in cell (1,1,1), for example, is

$$n\hat{\pi}_{111} = \frac{x_{1\cdot 1} x_{\cdot 11}}{x_{\cdot \cdot 1}} = \frac{2584 \cdot 342}{2884} = 306.42 \ .$$

The expected numbers in all the cells under H_2 are shown in Table 3.9.

TABLE 3.9. Expected numbers under H_2 for the data i Table 3.7.

A: Freezer	B: Social group	C: Ownership	
		Owner	Renter
Yes	I-II	306.42	89.78
	III	672.88	165.18
	IV	884.33	402.88
	V	720.37	420.15
No	I-II	35.58	66.21
	III	78.12	121.82
	IV	102.67	297.12
	V	83.63	309.85

This example can be used to illustrate the important point, that conditional independence given a third variable does not necessarily imply unconditional independence. As Figure 3.2 shows ownership of a freezer and social group are independent variables among owners as well as among renters taken separately. But if we study the **marginal distribution** of variables A and B shown in Table 3.10, they are no longer independent.

TABLE 3.10. The marginal contingency table of variables A and B.

Social group	Freezer	
	Yes	No
I-II	396	102
III	840	198
IV	1273	414
V	1153	381

The z-test statistic for independence between A and B computed from the numbers in Table 3.10 is

$$z = 16.36, df = 3.$$

Since $\chi^2_{0.95}(3) = 7.81$ *we must reject a hypothesis of independence (marginally) between social group and the possession of a freezer.*

As Example 3.2 demonstrated conditional independence does not necessarily imply marginal independence, that is A⊥B|C does not imply A⊥B.

3.5 Interpretation of the log-linear parameters

It is a consequence of the definition (3.2) of the log-linear parameters that the higher their values, the larger the value of the expected numbers for those cells which have subscripts in common with the parameter in question, and the lower their values, the smaller the value of the expected numbers. It thus follows that a negative value of a log-linear parameter means that the expected numbers are less that if the parameter in question is 0, while a positive value, on the other hand, means that the expected numbers become larger than if the parameter has the value 0. Finally since all log-linear parameters according to (3.3), (3.4) and (3.5) sum to 0 over all indices, their values illustrate the relative importance of the various interactions as compared to the baseline value 0.

As an example, if the value of τ^{AB}_{ij} is positive for a given combination of i and j, then the expected values in cells with A at level i and B at level j have larger values than if $\tau^{AB}_{ij} = 0$. Example 3.2 illustrates this.

EXAMPLE 3.2 (continued). *As we saw in section 3.4, the hypothesis* H_2 *can be accepted for the data in Table 3.6. Under* H_2 *the model contains the following parameters:* τ_{ij}^{AC}, τ_{jk}^{BC}, τ_i^A, τ_j^B *and* τ_k^C. *The ML-estimates for these parameters are shown below:*

$\hat{\tau}^{AC}_{ik}$	k = 1	2
i = 1	0.462	-0.462
2	-0.462	0.462

$\hat{\tau}^{BC}_{jk}$	k = 1	2
j = 1	0.119	-0.119
2	0.208	-0.208
3	-0.102	0.102
4	-0.225	0.225

$\hat{\tau}^{A}_{i}$	i = 1	2
	0.614	-0.614

$\hat{\tau}^{B}_{j}$	j = 1	2	3	4
	-0.794	-0.096	0.486	0.405

$\hat{\tau}^{C}_{k}$	k = 1	2
	0.032	-0.032

The interactions between A: Possession of a freezer and C: Ownership shows that more owners than renters have a freezer. The interactions between B: Social group and C: Ownership shows that owners are more often found in the high social groups, while renters are more often found in the lower social groups. The main effects $\hat{\tau}^{A}_{i}$ tell us that marginally, there are more Danes with a freezer in the household than without. According to the relative magnitude of the $\hat{\tau}^{B}_{j}$'s there are more Danes in the lower social groups, and finally the $\hat{\tau}^{C}_{j}$'s show that there are about the same number of owners and renters.

3.6 Choice of model

In discussion of contingency tables it is tradition to talk about models rather than hypotheses. In fact, for every hypothesis H_1 to H_7 there is a corresponding multinomial model (3.1) with cell probabilities defined by (3.2), without those log-linear parameters included which are specified as zero under the hypothesis. Under the hypothesis H_3, for example, all three-factor interactions and all two-factor interactions between A and B and between A and C are 0. This means that the model corresponding to H_3 is a multinomial distribution, where the cell probabilities π_{ijk} satisfy

$$\ln\left(\pi_{ijk}\right) = \tau_0^* + \tau_i^A + \tau_j^B + \tau_k^C + \tau_{jk}^{BC}.$$

This model is simpler than the saturated model (3.2), both in the sense that there are fewer parameters to keep track of, and in the sense that it has a simpler interpretation. The purpose of a statistical analysis of a contingency table is basically to describe the data in the table by a statistical model as simple as possible. Testing the hypotheses we considered in section 3.3 is thus part of a well structured statistical analysis.

Often one does not have a specific model in mind prior to the analysis. Confronted with the data the only possible approach is to try different models and evaluate their relative merits. For three-way contingency tables (in contrast to higher order contingency tables) it is possible to list all possible hypotheses and the corresponding models, together with the z-test statistics (3.19) for goodness of fit of the model. If we omit all hypotheses connected with uniform distribution over categories, that is the generic types H_4^*, H_5, H_6 and H_7, the list is even short enough to be displayed in a convenient table. From such a table, it is then in many cases possible to choose directly a suitable model to describe the data.

EXAMPLE 3.2 (continued). *Table 3.11 shows for the data in Table 3.6 the test statistics for all possible hypotheses except hypotheses of uniform distribution over categories*

TABLE 3.11. Test statistics for all possible hypotheses except hypotheses of uniform distribution over categories for the data in Table 3.6.

Hypothesis	Sufficient marginals	z(H)	df	Level of signif.
H_1	AB,AC,BC	5.97	3	0.113
(H_2)	AB,AC	113.27	6	0.000
(H_2)	AB,BC	644.70	4	0.000
H_2	AC,BC	7.38	6	0.287
(H_3)	AB,C	766.95	7	0.000
(H_3)	AC,B	129.62	9	0.000
H_3	BC,A	661.06	7	0.000
H_4	A,B,C	783.30	10	0.000

Table 3.11 shows that those hypotheses we chose to consider in Table 3.7 were indeed well chosen, since all hypotheses obtained from the generic types H_2 and H_3 by interchange of letters (indicated by parentheses in Table 3.11) obviously can not be accepted at any reasonable level.

There is a common and basic problem connected with using tables like Table 3.11 where many possible models are compared, especially for multi-way contingency tables, where the number of possible hypotheses to test - and thus models to compare - can be unmanageably large. The problem is that if we test all hypotheses at the same level, for example $\alpha = 0.05$, the level for the simultaneous test becomes much larger. By test level, we shall in this connection mean the probability of (incorrectly) rejecting a true hypothesis. In Table 3.11 there are 8 test statistics z(H). Even in a most favorable situation, where all test statistics are independent - which they definitely are not in this case - the simultaneous test level will with level $\alpha = 0.05$ in each individual test, be

$$1 - (1 - 0.05)^8 = 0.337 .$$

This, as we shall see below, is the probability that at least one true hypothesis is rejected. If we thus perform 8 tests, each with probability 0.05 of rejecting a true hypothesis, then in the end the probability is more than 1/3 that we have rejected at least one true hypothesis.

In order to show that if each test is performed at level 0.05, then the simultaneous test level is 0.337, let the test statistic for the j'th test be Z_j and let

$$P(Z_j \geq c_j) = 0.05 ,$$

where c_j is the critical value, for example $\chi^2_{0.95}(3)$ for Z_1. Then the probability of incorrectly rejecting at least one true hypothesis is

$$P(Z_j \geq c_j \text{ for at least one } j) = 1 - P(Z_j < c_j \text{ for all } j) .$$

If the Z_j's are independent it follows from the rules of probability calculus that

$$P(Z_j < c_j \text{ for all } j) = \prod_{j=1}^{m} P(Z_j < c_j)$$

and hence

$$P(Z_j \geq c_j \text{ for at least one } j) = 1 - (1 - \alpha)^m ,$$

if each test has level α and there are m tests. This is the formula we used above with m=8 and $\alpha = 0.05$.

To cope with simultaneous test situations, one possibility is to use the so-called **Bonferroni procedure**. This test procedure guaranties that the simultaneous test level for several tests **does not exceed** α. In its original form the Bonferroni procedure simply prescribed to use the test level α/m for each of m hypotheses tested simultaneously. Unfortunately the cost of this guaranty is that the actual simultaneous test level usually becomes considerably lower than α. Holm (cf. bibliographical notes) has improved the Bonferroni procedure, by developing a **sequential Bonferroni procedure**, for which the simultaneous test level is kept much closer to the intended level α, while still guaranteeing a simultaneous level of at most α.

The sequential Bonferroni procedure is as follows:

(i) *Ordering.* The m test statistics $Z_1, \ldots, Z_m$ are ordered according to increasing level of significance, that is if $z_1, \ldots, z_m$ are the observed values of the test statistics, then the ordered Z's, $Z_{(1)}, Z_{(2)}, \ldots, Z_{(m)}$ satisfy

$$P(Z_{(1)} \geq z_{(1)}) \leq \ldots \leq P(Z_{(m)} \geq z_{(m)}) \ .$$

(ii) *Step 1.* The hypothesis connected with $Z_{(1)}$ is treated first. If

$$P(Z_{(1)} \geq z_{(1)}) > \frac{\alpha}{m}$$

we accept all m hypotheses and the procedure stops at step 1. If on the other hand

$$P(Z_{(1)} \geq z_{(1)}) \leq \frac{\alpha}{m}$$

then $H_{(1)}$ is rejected and the procedure continues.

(iii) *Step 2.* Now the hypothesis connected with $Z_{(2)}$ is treated. If

$$P(Z_{(2)} \geq z_{(2)}) > \frac{\alpha}{m-1}$$

all subsequent hypotheses $H_{(2)}, \ldots, H_{(m)}$, i.e those connected with $Z_{(2)}, \ldots, Z_{(m)}$, are accepted and the procedure stops. If

$$P(Z_{(2)} \geq z_{(2)}) \leq \frac{\alpha}{m-1}$$

we reject $H_{(2)}$ and continue the procedure.

(iv) *Steps 3 to m-1.* The procedure continues in the same way with $Z_{(3)}, Z_{(4)}$ and so on, until we no longer reject hypotheses. In step 3 we use $\alpha/(m-2)$ in the inequalities, in step 4 $\alpha/(m-3)$, etc.

(v) *Step m.* If all hypotheses, except the one connected with $Z_{(m)}$, has been rejected, that is we have carried out m-1 steps of the procedure, the last step is to accept $H_{(m)}$ if

$$P(Z_{(m)} \geq z_{(m)}) > \alpha$$

and reject if

$$P(Z_{(m)} \geq z_{(m)}) \leq \alpha \ .$$

The best way to illustrate the use of the procedure is to apply it.

EXAMPLE 3.3. *Opinion on sports jointly with the opposite sex.*
The data in Table 3.12 is from an investigation of Danish school children. Among other things, they were asked about their views concerning sports at school. The three categorical variables considered here are:

A: *Opinion on sports jointly with the opposite sex.*
B: *School category.*
C: *Sex.*

TABLE 3.12. 625 Danish school children cross-classified according to Opinion on sport jointly with the opposite sex, School category and Sex.

A: Opinion on joint sport	B: School category	C: Sex Boys	Girls
Very good idea	Vocational school	27	13
	Commercial school	15	40
	High school	31	103
Good idea	Vocational school	31	10
	Commercial school	21	43
	High school	51	67
Bad idea	Vocational school	12	6
	Commercial school	23	18
	High school	38	29
Very bad idea	Vocational school	2	3
	Commercial school	7	4
	High school	14	17

Source: Scholin, B. (1989): The activities in and attitudes towards sport among Danish school children, age 16-19. (In Danish). Danish School for Higher Physical Education. (Danmarks Højskole for Legemsøvelser.)

Note: Between the age of 16 and 18 Danish school children can choose between three types of high schools (or junior colleges); one leading to vocational jobs, like carpenters, automechanics or electricians, one leading to commercial jobs in stores or in offices, and finally the traditional high school leading to further theoretical education.

If we use the data in Table 3.12 to order the 8 hypotheses in Table 3.11, identified by their sufficient marginals, according to increasing value of the level of significance, we get Table 3.13, where also the value of $0.05/(9-j)$ for $j = 1,...,8$ is shown.

TABLE 3.13. Eight hypotheses for the data i Table 3.12 ordered according to increasing value of the level of significance.

	Hypothesis	Sufficient marginals	z(H)	df	Level of signif.	0.05 / (9-j)
j=1	$H_{(1)}$	AB,C	73.44	11	0.00000	0.00625
2	$H_{(2)}$	A,B,C	79.71	17	0.00000	0.00714
3	$H_{(3)}$	AB,AC	48.68	8	0.00000	0.00833
4	$H_{(4)}$	AC,B	54.95	14	0.00000	0.01000
5	$H_{(5)}$	AB,BC	39.81	9	0.00001	0.01250
6	$H_{(6)}$	BC,A	46.08	15	0.00005	0.01667
7	$H_{(7)}$	AC,BC	21.33	12	0.04581	0.02500
8	$H_{(8)}$	AB,AC,BC	11.50	6	0.07400	0.05000

If we follow the sequential Bonferroni procedure, we must reject the first 6 hypotheses and when in step seven 0.04581 > 0.025, we accept the remaining two hypotheses $H_{(7)}$ and $H_{(8)}$. Note here that $H_{(7)}$, with sufficient marginals AC and BC, is accepted at level 0.05, although the level of significance for a direct test is 0.046 < 0.05. This example thus shows that when testing several hypotheses one should not to be too strict with obeying the prescribed over-all level of significance.

In the process of choosing a model we can also use the sequential test statistic (3.28), where two nested hypotheses $H \subset H^*$ are compared. This means that instead of comparing the model corresponding to H to the saturated model by means of Z(H), we evaluate the goodness of fit of the model corresponding to H by comparing it to H^*, where H^* corresponds to a more complicated - although not the saturated - model by means of

$$Z(H|H^*) = Z(H) - Z(H^*) = 2 \sum_i \sum_j \sum_k X_{ijk} \left(\ln(n\hat{\pi}^*_{ijk}) - \ln(n\hat{\pi}_{ijk}) \right).$$

This is equivalent to testing if the model under hypothesis H describes the data as well as the model under the alternative hypothesis H^*.

EXAMPLE 3.3 (continued). *Suppose that for the data in Table 3.12 we have decided to test the fit of various models in such a way that we try in turn to omit the interactions $\tau^{ABC}_{ijk}, \tau^{AB}_{ij}, \tau^{AC}_{ik}$ and τ^{BC}_{jk} in that order. This means that a priori we regard it as most likely that the two-factor interaction between Opinion on joint sport with the other sex and School category can be omitted, and it is least likely that the two-factor interaction between Sex and School category can be omitted. This leads to the four **hierarchically ordered** hypotheses shown in Table 3.14. Also shown in the table are the observed values of Z(H) and Z(H|H^*).*

TABLE 3.14. Four hierarchically ordered hypotheses for the data in Table 3.12 shown with z(H) and z(H|H*), where H* is the previous hypothesis in the list.

H	Sufficient marginal	Omitted interaction	z(H)	df	Level of sign.	z(H|H$^\cdot$)	df	Level of sign.
H$_1$	AB,AC,BC	ABC	11.50	6	0.074	-		-
H$_2$	AC,BC	AB	21.33	12	0.046	9.83	6	0.132
H$_3$	BC,A	AC	46.08	15	0.000	24.75	3	0.000
H$_4$	A,B,C	BC	79.71	17	0.000	33.63	2	0.000

Table 3.14 shows that the three-factor interactions can be omitted and that also the two-factor interactions between variables A and B can be omitted. Both these conclusions are in agreement with the conclusion from the sequential Bonferroni procedure.

In order to illustrate that it occasionally happens that we want to test uniform distribution over categories when all the variables are independent, we return to Example 3.1.

EXAMPLE 3.1 (continued). *Table 3.15 shows the test statistics for a sequence of models applied to the data in Table 3.1.*

TABLE 3.15. Test statistics for selected hypotheses for the data in Table 3.1, where z(H|H*) is the sequential test statistic and H* the previous hypothesis in the list.

H	Sufficient marginals	Omitted interaction	z(H)	df	Level of signif.	z(H|H$^{*)}$)	df	Level of signif.
H$_1$	AB,AC,BC	ABC	6.78	4	0.148	-	-	-
H$_2$	AB,BC	AC	7.37	6	0.288	0.59	2	0.745
H$_3$	AB,C	BC	9.08	8	0.336	1.71	2	0.245
H$_4$	A,B,C	AB	13.61	12	0.326	4.53	4	0.339
H$_5$	A,B	C	13.93	13	0.379	0.32	1	0.572
H$_6$	A	B	17.16	15	0.309	3.23	2	0.199
H$_7$	-	A	620.90	17	0.000	603.75	2	0.000

For these data the independence hypothesis with sufficient marginals A,B,C is accepted, and we have the possibility, given independence, to check uniform distribution over categories. Table 3.15 shows that all variables have a uniform distribution over categories except variable A: Frequency of watching videos. The final model is thus a model with sufficient marginals A,B and C, but with uniform distribution over the categories of variables B and C, that is Age and Sex. The uniform distribution over Age and Sex means that the sample is representative of the population of school children age 7 to 15 since there is approximately the same number of boys and girls and the same number for each 3-year age interval in the population. The association diagram for the final model is shown in Figure 3.3.

FIGURE 3.3. Association diagram for a model with sufficient marginals A, B and C and uniform distribution over the categories of variables B and C,

3.7 Detection of model deviations

It often happens that an otherwise interesting model is rejected as a satisfactory fit to the data, in some cases clearly against the data collector's expectations. Such cases call for a closer inspection of the data. It may for example be that the model fits the data except for a few cells. The model is then rejected on the weight of the model deviations in these few cells. In case there is a natural explanation for the behavior in these few cells, one can give a more accurate summary of the search for a model, for example a statement like: "By and large the following simple model fits the data, except for a few - to be expected - model deviations connected with the following combinations of variable categories ....". The way to find out if we are in such a situation, is to use **standardized residuals**. A residual is the difference between the observed and the expected number in cell (ijk) under a given hypothesis H, that is

$$(x_{ijk} - n\hat{\pi}_{ijk}) ,$$

where $\hat{\pi}_{ijk}$ is the estimated cell probability under H. The standardized residuals are then

$$r_{ijk} = (x_{ijk} - n\hat{\pi}_{ijk})/\hat{\sigma}_{ijk} , \qquad (3.29)$$

where

$$\sigma^2_{ijk} = \text{var}[X_{ijk} - n\hat{\pi}_{ijk}] , \qquad (3.30)$$

and $\hat{\sigma}^2_{ijk}$ is an estimate of σ^2_{ijk}.

Since the model is a log-linear model, $\hat{\sigma}^2_{ijk}$ is obtained by inserting the estimates $\hat{n}_{ijk}$ under H in formula (2.46). The design matrix **W** and the inverse of **M** are not computed as a by-product of the iterative proportional fitting procedure and thus have to be computed separately. On the other hand, the calculations needed to estimate the standard errors of the ML-estimates for the log-linear parameters are easy. Surprisingly many major statistical program packages still do not pro-vide standardized residuals, although they compute the standard errors for the ML-estimates and often also the elements of **H**.

For all the decomposable models, where we have closed forms for the expected values, it can be shown that there are also a closed form for $\hat{\sigma}^2_{ijk}$. The required expression is obtained by noting from Table 3.5 that all expected values for decomposable models for a certain b have the form

$$\hat{\mu}_{ijk} = \frac{x_{[1]} \cdots x_{[b]}}{n_{[2]} \cdots n_{[b]}},$$

where $x_{[v]}$ and $n_{[v]}$ are totals of the contingency table. It was shown by Haberman (see biographical notes) that the values of $\hat{\sigma}^2_{ijk}$ are the given by

$$\hat{\sigma}^2_{ijk} = \hat{\mu}_{ijk}\left[1 - \hat{\mu}_{ijk}\left(\sum_{v=1}^{b} \frac{1}{x_{[v]}} - \sum_{v=2}^{b} \frac{1}{n_{[v]}}\right)\right]. \quad (3.31)$$

The values of (3.31) for hypotheses H_2, H_3 and H_4 are shown in Table 3.16.

TABLE 3.16. Estimated variances of the residuals for hypotheses H_2, H_3 and H_3.

Hypothesis	Estimated variance $\hat{\sigma}^2_{ijk}$ of the residuals
H_2	$\hat{\mu}_{ijk}(1 - x_{i.k}/x_{..k})(1 - x_{.jk}/x_{..k})$
H_3	$\hat{\mu}_{ijk}(1 - x_{.jk}/n)(1 - x_{i..}/n)$
H_4	$\hat{\mu}_{ijk}(1 - x_{i..}x_{.j.}/n^2 - x_{i..}x_{..k}/n^2 - x_{.j.}x_{..k}/n^2 + 2x_{i..}x_{.j.}x_{..k}/n^3)$

There is a standardized residual for each cell in the contingency table and r_{ijk} is approximately distributed as a standard normal random variable. Model deviations are, therefore, most likely to be found in cells for which

$$|r_{ijk}| > 2.$$

Quantities like the standardized residuals (often in program packages called "adjusted" residuals) are known in modern statistics as **diagnostics**.

It is important to keep in mind that any of the generic types of hypotheses assumes that **all** the interactions of a certain type are 0. But we may face situations where **almost all** interactions of a certain type are zero, **expect for a few large ones**. The hypothesis is then rejected on the weight of these few large interactions. In case there is a natural explanation for these few interactions to be large, one can then give a more accurate summary of the search for a model, for example: "By and large the following simple model fits the data, except for a few large interactions between the following variables ....". The diagnostics for detecting such situations, are the **standardized parameter estimates**, for example

$$\hat{\omega}_{ij}^{AB} = \frac{\hat{\tau}_{ij}^{AB}}{\hat{\sigma}(\tau_{ij}^{AB})},$$

for τ_{ij}^{AB} where

$$\sigma^2(\tau_{ij}^{AB}) = \text{var}\left[\hat{\tau}_{ij}^{AB}\right],$$

and $\hat{\sigma}^2(\tau_{ij}^{AB})$ is an estimate of $\sigma^2(\tau_{ij}^{AB})$.

Also the standardized parameter estimates are approximately distributed as standard normal random variables. They can, therefore, be used to detect combinations of levels i and j for which

$$\left|\hat{\omega}_{ij}^{AB}\right| > 2,$$

indicating cells where the model fit is poor.

EXAMPLE 3.4. *Broken marriages.*
We have for this example selected the following variables from the Danish Welfare Study (cf. Example 1.1 and 3.2):

   A: *Social group*
   B: *Sex*
   C: *Broken marriage or not*

The cross-classification of these categorical variables is shown in Table 3.17.

TABLE 3.17. The sample from the Danish Welfare Study cross-classified according to Social group, Sex and Broken marriage or not.

A: Social group	B: Sex	C: Broken marriage	
		Yes	No
I	Men	14	102
	Women	12	25
II	Men	39	15
	Women	23	79
III	Men	42	292
	Women	37	151
IV	Men	79	293
	Women	102	557
V	Men	66	261
	Women	58	321

Source: The data base of the Danish Welfare Study. Cf. Example 3.2.

Note: Because of more modern forms of living, for example in Denmark in 1974, "marriage" is a common term for both a formal marriage and a permanent relationship.

Table 3.18 shows the test statistics for all the hypotheses of generic types H_1, H_2, H_3 and H_4.

TABLE 3.18. Test statistics for 8 hypotheses and the data in Table 3.18.

Hypothesis	Sufficient marginals	z(H)	df	Level of signif.
H_1	AB,AC,BC	19.89	4	0.001
H_2	AB,AC	20.38	5	0.001
H_2	AB,BC	24.78	8	0.002
H_2	AC,BC	215.67	8	0.000
H_3	AB,C	25.18	9	0.003
H_3	AC,B	216.07	9	0.000
H_3	BC,A	220.47	12	0.000
H_4	A B C	220.88	13	0.000

Obviously none of the hypotheses can be accepted at a reasonable test level, which means that no simple model with an interesting interpretation can be found based on this data set. There is, however, a pattern in the sizes of the observed test statistics and the levels of significance. This phenomenon is further illustrated by

THREE-WAY CONTINGENCY TABLES 77

the sequential tests in Table 3.19, where the models are compared with the model having sufficient marginals AB, AC, BC, that is hypothesis H_1.

TABLE 3.19. Sequential test statistics for some of the hypotheses in table 3.18.

Hypothesis	Sufficient marginals	z(H|H$_1$)	df	Level of significance
H_2	AB, AC	0.49	1	0.483
H_2	AB, BC	4.89	4	0.300
H_3	AB, C	5.29	5	0.382

A model with sufficient marginals AB, C thus describes the data in the contingency table just as well as a model with sufficient marginals AB, AC, BC. The model under H_3 is interesting, since it has the interpretation $C \perp A,B$ and hence the association diagram shown in Figure 3.4.

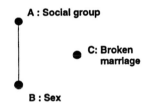

FIGURE 3.4. Association diagram for the model AB, C.

A model with interpretation $C \perp A,B$ means that whether one has a broken marriage or not is independent of sex and social group; that is broken marriages occur equally frequently among men and women and the probability of having a broken marriage is the same in all five social groups. Unfortunately H_3 does not fit the data, but this might be due to a few deviations from the expected pattern under H_3 in the cells. In order to explore this possibility, we consider the standardized residuals under H_3, shown in Table 3.20.

TABLE 3.20. Standardized residuals for the data in Table 3.18 under H_3.

A: Social group	B: Sex	C: Broken marriage Yes	No
I	Men	-1.56	1.56
	Women	2.42	-2.42
II	Men	1.16	-1.16
	Women	1.38	-1.38
III	Men	-2.51	2.51
	Women	0.83	-0.83
IV	Men	2.07	-2.07
	Women	-1.54	1.54
V	Men	1.39	-1.39
	Women	-1.19	1.19

As the table shows, it is in fact only 4 of the 20 cells that account for the model deviations, in the sense that the residuals are substantially larger numericallythan 2. These four cells are interesting. In social group I, there are more women that we should expect who have a broken marriage. Virtually all women in social group I in Denmark in 1974 had academic degrees because very few women had top positions in the private sector. The fact that so many broken marriages were found among women in social group I, therefore, gave rise to the headline "Academic women in top" for the article where these data were first presented. In social group III, on the other hand, men seem to have a broken marriage less often than women. In 1974 men in social group III were to a large extent owners of small businesses such as grocery stores, carpenter firms or sanitary repair firms. In such businesses the wife was often part of the firm as bookkeeper, giving a hand in the store, or taking the telephone calls. One explanation could therefore be that a couple would hesitate to divorce, because both partners are needed to secure the income. The reason for women with an academic degree to divorce more often can have several explanations. One might be more economical freedom.

3.8 Bibliographical notes

The classical book on log-linear models for contingency tables is Bishop, Fienberg and Holland (1975), which contains all basic results including the asymptotic results for estimators and tests. For general reference readers are referred to the recent monographs covering log-linear models for contingency tables by Christensen (1990), Agresti (1990) and Andersen (1990). General references for graphical models are Whittager (1990) and Edwards (1995).

The log-linear parametrization for three-way tables was introduced by Birch (1963).

THREE-WAY CONTINGENCY TABLES 79

The theory of log-linear models for contingency tables was developed by Goodman (1968). The major papers by Goodman are collected in the monograph by Goodman (1978). A wealth of statistical important results, including the asymptotic theory for estimators and tests, was given by Haberman (1974) on a high mathematical level. An easier accessible account is given in the two volume book Haberman (1978), (1979).

Graphical representations for associations in log-linear models were introduced by Goodman (1972) and (1973). A general theory for graphical models was developed by Darroch, Lauritzen and Speed (1980), cf. also Lauritzen (1996).

The sequential Bonferroni procedure was introduced by Holm (1979) and further developed by Schaffer (1986). Standardized residuals for the multinomial distribution were introduced by Rao (1973). For contingency tables they were fully developed by Haberman (1974).

3.9 Exercises

[In some of these exercises it is necessary to apply statistical packages like SPSS, BMDP or SAS to compute test statistics and parameter estimates. For readers without access to such packages a number of key test statistics and parameter estimates are given for selected exercises in the Appendix.]

3.1 Consider a 2×2×2 contingency table where all the expected numbers are equal and the total is n = 480.

(a) Show that all log-linear parameters except τ_0 are zero.

(b) Estimate both τ_0 and τ_0^*.

3.2 Consider a 2×2×2 contingency table with observed numbers

		C = 1	2
A = 1	B = 1	71	18
	2	50	38
2	B = 1	18	31
	2	21	91

(a) Estimate the log-linear parameters τ_{ij}^{AB} for all i and j.

(b) Estimate the main effects τ_j^B for all j.

3.3 For a 3×2×2 contingency table let the expected values be

			C = 1	2
A = 1	B = 1		40.5	28.7
	2		34.7	36.5
2	B = 1		64.6	31.4
	2		70.4	41.5
3	B = 1		56.7	32.7
	2		13.5	7.8

(a) Compute the values of the interactions τ_{jk}^{BC} for all j and k. [Hint: Use an analogue to the formula just after Equation (3.6) in Section 3.1.]

(b) Compute the values of the main effects τ_k^C.

3.4 Consider a 4×5×3 contingency table.

(a) Calculate the number of unconstrained log-linear parameters for all interactions and main effects (except τ_0).

(b) Check that the numbers you found in (a) add up to 59. Why is the total 59?

3.5 Consider the hypotheses H_1 through H_4 given by

H_1: $\tau_{ijk}^{ABC} = 0$ for all i, j and k
H_2: $\tau_{ijk}^{ABC} = \tau_{jk}^{BC} = 0$ for all i, j and k
H_3: $\tau_{ijk}^{ABC} = \tau_{jk}^{BC} = \tau_{ik}^{AC} = 0$ for all i, j and k
H_4^*: $\tau_{ijk}^{ABC} = \tau_{jk}^{BC} = \tau_{ik}^{AC} = \tau_k^C = 0$ for all i, j and k
H_4: $\tau_{ijk}^{ABC} = \tau_{jk}^{BC} = \tau_{ik}^{AC} = \tau_{ij}^{AB} = 0$ for all i, j and k

rather than those given at the start of Section 3.2.

(a) Write down the interpretations of H_2 through H_4 in the same manner as in Theorem 3.1.

(b) Give the exact expressions for π_{ijk} under H_2 through H_4 in terms of the marginals of π_{ijk}.

(c) Draw the association diagrams of H_2 through H_4.

3.6 Reconsider the data in exercise 3.2.
(a) Estimate the expected numbers under the model AC, BC.

(b) Draw the association diagram for the model in (a).

Under the model AB, AC, BC the estimated expected numbers are

			C = 1	2
A = 1	B = 1		70.3	18.7
	2		50.7	37.3
2	B = 1		18.7	30.3
	2		20.3	91.7

(c) Carry out steps 1 through 3 of iteration 1 for the iterative proportional fitting procedure, as described in Section 3.3, Example 3.1 and compare the results obtained with the expected values shown above.

(d) Test the fit of both model AC, BC and model AB, AC, BC.

3.7 The table below shows the Non-response (variable A) cross-classified with Sex (variable B) and Residence (variable C) in a survey conducted by the Danish National Institute of Social Research.

		C: Sex	
A: Response	B: Residence	Male	Female
	Copenhagen	306	264
Yes	Cities	609	627
	Countryside	978	947
	Copenhagen	49	76
No	Cities	77	79
	Countryside	103	114

(a) Determine a suitable model which fits the data.

(b) Draw the association diagram for the chosen model and interpret the model.

(c) Compute the ML-estimates of any 2-factor interactions in the model based on the expected numbers under the model. What do they tell you about the way the variables interact?

3.8 The Swedish Traffic Authorities counted the number of persons killed in the traffic in trial periods of 18 weeks in 1961 and 18 weeks in 1962. In 9 weeks of each of these periods a speed limit of 90 km per hour was imposed, while in the remaining 9 weeks of the periods there were no speed limits. The number of killed were cross-classified according to A: Speed limit or not, B: Year and C: Type of road.

		C: Type of road	
A: Speed limit	B: Year	Main	Secondary
90 km/hour	1961	8	42
	1962	11	37
Free	1961	57	106
	1962	45	69

(a) Use direct or sequential tests to determine a simple model which fits the data.

(b) Make an interpretation of the selected model based on an association diagram.

(c) Compute the z(H)-test statistics for the following models in turn and apply the sequential Bonferroni procedure.

AB, AC, BC
AC, BC
AC, B
AC
A, B, C
A, C
C

3.9 Radio Denmark conducted in 1985 a survey of the interest among TV-viewers concerning the length (variable A) of the Saturday afternoon TV-program called "Saturday Sports" featuring various sports events. The Preferences (variable A) as regards the length of the program were cross-classified with Age (variable B) and Sex (variable C). The results were

A: Preference		C: Sex	
of length	B: Age	Male	Female
Less than	Under 40	65	77
2 hours	Over 40	81	80
2½ to 3½	Under 40	63	39
hours	Over 40	50	38
4 hours	Under 40	59	32
or more	Over 40	30	6

(a) Find a suitable simple model which fits the data.

(b) Explain how you have used test statistics to select the model.

(c) Draw the association diagram of the selected model and give an interpretation of the model.

(d) Compute the expected values under the chosen model.

3.10 Twins can be either Monocygotes (coming from one egg) or Dicygotes (coming from different eggs). Monocygotes can be expected to have identical genetical characteristics. Twin studies are, therefore, important in genetic research. In order to test the hypothesis that alcohol abuse is hereditary information was collected in Finland and Sweden on alcohol abuse of the second twin, if alcohol abuse was already established for one of the twins. The table below shows the results.

		C: Alcohol abuse	
A: Country	B: Twin type	Both	Only one
Finland	Monocygote	159	1102
	Dicygote	220	2696
Sweden	Monocygote	132	1171
	Dicygote	165	1756

(a) Select a simple model, which fits the data and draw its association diagram.

(b) What does the association diagram tell you about alcohol abuse being hereditary?

(c) Are there differences in alcohol abuse between the two involved Scandinavian countries?

3.11 Derive the expressions in Table 3.16 from Equation (3.24).

3.12 Compute the standardized residuals for models (a) AB, BC, (b) A, BC and (c) A, B, C for the data in Exercise 3.10.

Chapter 4

Multi-dimensional contingency tables

4.1 The log-linear model

An m-dimensional contingency table is an m-dimensional array of observed counts

$$x_{i_1 i_2 \ldots i_m}$$

with index $i_1,\ldots,i_m$, which is the result of a cross-classification of m categorical variables. When I_j is the number of categories for variable number j, the range of i_j is

$$i_j = 1,\ldots,I_j .$$

The **dimension** of the contingency table is thus $I_1 \cdot \ldots \cdot I_m$.

We shall use I, J, K, and L, instead of I_1, I_2, I_3 and I_4 when the contingency table is of dimension 2, 3 and 4, and use the capital letters A,B,C and D as generic names for the categorical variables.

As statistical model for an m-dimensional contingency table we shall use a **multinomial model**, such that the vector of random variables

$$(X_{11\ldots 1},\ldots,X_{I_1 I_2 \ldots I_m})$$

connected with the observed counts

$$(x_{11\ldots 1},\ldots,x_{I_1 I_2 \ldots I_m})$$

has the distribution

$$(X_{11\ldots 1},\ldots,X_{I_1 I_2 \ldots I_m}) \sim M\left(n;\pi_{11\ldots 1},\ldots,\pi_{I_1 \ldots I_m}\right) . \qquad (4.1)$$

This model can also arise from a **Poisson model**, where the X's are independent Poisson distributed, cf. the remark in section 2.6 and formula (2.28).

The multinomial model for a multi-dimensional contingency table is called a **log-linear model**, if the mean values

$$\mu_{i_1\ldots i_m} = E\left[X_{i_1\ldots i_m}\right] = n\pi_{i_1\ldots i_m}$$

have the parametric structure

$$\begin{aligned}\ln\mu_{i_1\ldots i_m} &= \tau^{AB\ldots S}_{i_1\ldots i_m} + \ldots + \tau^{ABC}_{i_1\ldots i_3} + \ldots + \tau^{QRS}_{i_{m-2}i_{m-1}i_m} \\ &+ \tau^{AB}_{i_1 i_2} + \ldots + \tau^{RS}_{i_{m-1}i_m} + \tau^{A}_{i_1} + \ldots + \tau^{S}_{i_m} + \tau_0 \; .\end{aligned} \qquad (4.2)$$

If a τ-parameter has more than one subscript it is called an **interaction**, while a τ with only one subscript is called a **main effect**. With N subscripts an interaction is called an N-factor interaction. Thus with $i_1=i$, $i_2=j$, $i_3=k$ and $i_4=l$, τ^{ABCD}_{ijkl} is the 4-factor interaction between variables A, B, C and D, while τ^{AB}_{ij} is the 2-factor interaction between variables A and B, and τ^{A}_{i} is the main effect for variable A. As in section 3.1 (Equations (3.4) to (3.6)), the interactions and main effects are normalized so that they sum to zero over all indices, for example

$$\sum_{i_1} \tau^{ABC}_{i_1 i_2 i_3} = \sum_{i_2} \tau^{ABC}_{i_1 i_2 i_3} = \sum_{i_3} \tau^{ABC}_{i_1 i_2 i_3} = 0.$$

Since a contingency table is a cross-classification of m categorical variables, a very common application is to describe the result of a survey. If a random sample of n individuals is drawn from a population and for each individual we have observed the response on m questions in a questionnaire, all with a finite number of response categories, the cell count in cell $(i_1, \ldots ,i_m)$ is the number of individuals, who have responded in category i_1 on question A, in i_2 on question B, etc. If all n individuals have answered the questions independently, the multinomial model (4.1) is obviously the correct model. Alternatively to saying that question A is answered in category i_1, we may say that variable A is observed at level i_1, the latter expression being the most general. Hence the cell probabilities in (4.1) are

$$\pi_{i_1\ldots i_m} = P(A \text{ is observed at level } i_1, B \text{ at level } i_2, \ldots, S \text{ at level } i_m) \; .$$

The log-linear model (4.2) may seem complicated at first sight, but is in fact a very convenient way of keeping track of important hypotheses. Matters are also considerably simplified by the fact, as will be shown below, that we seldom need to write the full expression or even estimate the many interactions that are part of a typical model.

That model (4.2) is a log-linear model, as introduced in section 2.9, is readily seen from the defining Equation (2.39) for a log-linear model. The design matrix is on the other hand rather complicated because of the many constraints imposed by all subscripts adding to zero. Since the model is log-linear any hypothesis, defined in terms of interactions or main effects, can, however, be tested by Z-test statistics of the form

$$Z(H) = 2 \sum_{i_1=1}^{I_1} \cdots \sum_{i_m=1}^{I_m} X_{i_1 \cdots i_m} \left(\ln X_{i_1 \cdots i_m} - \ln \hat{\mu}_{i_1 \cdots i_m} \right), \qquad (4.3)$$

where H is a hypothesis specified by setting a certain subset of log-linear parameters equal to 0. In (4.3)

$$\hat{\mu}_{i_1 \cdots i_m} = n \hat{\pi}_{i_1 \cdots i_m},$$

where the $\hat{\pi}$'s are the estimated cell probabilities.

The fact that the test statistic (4.3) depends only on the expected mean values under H has important consequences for the statistical analysis of contingency tables. The iterative proportional fitting method results in estimated expected values rather than ML-estimates of the log-linear parameters. Hence we do not need additional computations in order to calculate the value of Z(H) by Equation (4.3).

The asymptotic distribution of Z(H) is a χ^2-distribution with df(H) degrees of freedom, where df(H) is the number of unconstrained τ's, which are 0 under H.

The significance level of H is evaluated by

$$p = P\big(Z(H) \geq z(H)\big) \approx P\big(Q \geq z(H)\big) \qquad (4.4)$$

where z(H) is the observed value of Z(H), and

$$Q \sim \chi^2(df(H)).$$

By a sequential test, we can evaluate the goodness of fit of the data under the model corresponding to H against the model corresponding to H^*, where H is a model with more log-linear parameters assumed to be zero than under H^*, i.e. $H \subset H^*$. For a sequential test, the test statistic is

$$Z(H \mid H^*) = Z(H) - Z(H^*) \qquad (4.5)$$

with level of significance

$$p = P\big(Z(H|H^*) \geq z(H|H^*)\big) \approx P\big(Q \geq z(H|H^*)\big) \qquad (4.6)$$

where $Q \sim \chi^2(df(H) - df(H^*))$. As we recall, if the value of (4.6) is above a certain test level, for example 0.05, we accept H as compared with H^*. This means that the model corresponding to H describes the date as well as the model corresponding to H^*. Since the model corresponding to H is the simpler one, we would then prefer this model as a description of the data.

MULTI-DIMENSIONAL CONTINGENCY TABLES

For contingency tables of high dimensions, there are many hypotheses and models to track. This is a consequence of model (4.2) being a **saturated model**, in the sense that it is equivalent to a multinomial model with no constraints on the cell probabilities. For three-dimensional tables, which we treated separately in chapter 3, there are a limited number of possible hypotheses and models, such that in principle one can take a look at the fit of all the possible models. This is not the case for higher order tables. Hence it becomes essential that we develop strategies to determine which models to inspect during the statistical analysis. Already for 4-way contingency tables there are so many possible models that it is difficult to set up a reasonable listing order for the hypotheses under consideration. It is, on the other hand, possible to make a classification of special generic types of hypotheses. This facilitates the interpretation of a model, once found satisfactory by a goodness of fit test.

The 4-way contingency table is attractive as an illustration of multiple contingency tables, because, on the one hand, the number of possible models to consider is rich enough to illustrate the problem of model search strategies and, on the other hand, the number of generic or typical models is limited, making a description of basic principles possible.

4.2 Classification and interpretation of log-linear models

We shall increasingly talk about **models** rather than hypotheses. As we have seen, to every hypothesis H, defined as a certain subset of the log-linear parameters being 0, there is a corresponding model, where the log-mean value of the counts is defined as the sum of those log-linear parameters not assumed to be 0 under the hypothesis.

It is important to repeat that for 4-way tables, we use the subscripts i, j, k and l rather than i_1, i_2, i_3 and i_4. For a 4-way contingency table the observed counts are thus denoted

$$x_{ijkl}, i = 1,...,I, j = 1,...,J, k = 1,...,K, l = 1,...,L$$

and the saturated log-linear model is

$$\ln \mu_{ijkl} = \tau_{ijkl}^{ABCD} + \tau_{ijk}^{ABC} + ... + \tau_{jkl}^{BCD}$$
$$+ \tau_{ij}^{AB} + ... + \tau_{kl}^{CD} + \tau_i^A + ... + \tau_l^D + \tau_0 ,$$

(4.7)

where $\mu_{ijkl} = E[X_{ijkl}]$.

As an example of a possible model consider the model corresponding to the hypothesis

$$H_1 : \tau^{ABCD}_{ijkl} = \tau^{ABC}_{ijk} = \tau^{ABD}_{ijl} = 0 \ . \qquad (4.8)$$

for all i, j, k and l. The model corresponding to H_1 is uniquely defined by the sufficient marginals

$$ACD, BCD, AB \ . \qquad (4.9)$$

To see this, we recall that all likelihood equations have the form

$$\text{observed sufficient marginal} = \text{expected sufficient marginal}$$

for all sufficient marginals corresponding to interactions not being zero under the model. The equations corresponding to the sufficient marginals ACD and BCD are

$$x_{i.kl} = E[X_{i.kl}] \ , \text{ for all i, k and l} \qquad (4.10)$$

and

$$x_{.jkl} = E[X_{.jkl}] \ , \text{ for all j, k and l} \ . \qquad (4.11)$$

In addition there are a number of likelihood equations corresponding to two-factor interactions and main effects. The majority of these will follow automatically from (4.10) and (4.11), for example

$$x_{i.k.} = E[X_{i.k.}] \ ,$$

which follows from (4.10) by summation over l. Only one equation corresponding to a 2-factor interaction can not be obtained by summation over either (4.10) or (4.11), namely

$$x_{ij..} = E[X_{ij..}] \ , \text{ for all i and j.} \qquad (4.12)$$

It follows that the sufficient marginals for the model corresponding to the hypothesis (4.8) are those on the left hand sides in Equations (4.10), (4.11) and (4.12), or in symbolic form: ACD, BCD, AB.

It is now easy to recognize the general pattern. For a given hypothesis, we

(a) write down all sufficient marginals corresponding to interactions of the highest dimension, which are not zero under the hypothesis.

(b) add all non-zero sufficient marginals of a lower dimension, which are not obtainable from the marginals in (a).

As an example we apply this method of selecting sufficient marginals to the log-linear model under the hypothesis

$$H_2 : \tau^{ABCD}_{ijkl} = \tau^{ABC}_{ijk} = \tau^{ABD}_{ijl} = \tau^{ACD}_{ikl} = 0 \ . \qquad (4.13)$$

Under (a) we must include the sufficient marginal $x_{.jkl}$, or in symbolic form BCD. Among the 2-factor interactions those marginals corresponding to BC, BD and CD can be derived from $x_{.jkl}$, but this is not the case for the 2-factor interactions corresponding to AC, AB and AD. If one does not want to check the many subscripts, an easy rule is the following: A two-factor interaction must be added, if the corresponding letter combination is not a subset of the letter combination for a 3-factor interaction, not assumed to be zero. Thus AC must be added, because the letters A and C are not a subset of the three letter group BCD. The model is accordingly determined by the sufficient marginals

$$BCD, AC, AB, AD \ . \qquad (4.14)$$

In this example we do not need to add sufficient marginals corresponding to main effects, since all letters are represented in (4.14).

In general the rule is as follows:

A lower dimension interaction between variables must be added, if the corresponding letter combination is not a subset of the letter combination for a higher dimension interaction, not assumed to be zero under the model.

Consider as an application the model corresponding to the hypothesis

$$H_3 : \text{All 3-factor and 4-factor interactions are 0}$$

$$\text{and } \tau^{AB}_{ij} = \tau^{AC}_{ik} = \tau^{AD}_{il} = 0 \ . \qquad (4.15)$$

In this case we first write down the interactions of highest dimension with sufficient marginals, which are not assumed to be zero, namely BC, BD, CD. In this case we have, however, to add the marginal $x_{i...}$, since the letter A is not represented in any of the included 2-factor interactions. Hence the sufficient marginals for the model are

$$BC, BD, CD, A \ .$$

The models corresponding to the hypotheses H_1, H_2 and H_3 are all examples of **hierarchical models**. A model is hierarchical if no likelihood equation corresponding to a sufficient marginal can be derived by summation over other likelihood equations, which corresponds to sufficient marginals defining the model. A model is therefore hierarchical, if it is uniquely determined by its sufficient marginals. The hypothesis

$$H_4 : \tau_{ijkl}^{ABCD} = \tau_{ijk}^{ABC} = \tau_{ikl}^{ACD} = \tau_{kl}^{CD} = 0$$

does not, according to this definition, correspond to a hierarchical model, since all τ_{ikl}^{ACD}'s are assumed to be 0, and the letter combination CD is a subset of ACD. Hence the likelihood equation $E[X_{..kl}] = x_{..kl}$ corresponding to the sufficient marginal CD can be derived from the likelihood equation corresponding to ACD. By our rule H_4 is uniquely determined by the sufficient marginals

$$ABD, BCD, AB ,$$

which give no indication of the fact that we have assumed that $\tau_{kl}^{CD} = 0$. In the rest of this chapter we only consider hierarchical hypothesis.

Note: The reason why non-hierarchical models are complicated to handle is primarily that the normalization and hence the definition of lower order interactions depends on the normalization of higher order interactions and accordingly which higher order interactions are included in the model.

As yet another example consider the model with sufficient marginals

$$ABC, AD, BD$$

corresponding to the hypothesis

$$H_5 : \tau_{ijkl}^{ABCD} = \tau_{ikl}^{ACD} = \tau_{ijl}^{ABD} = \tau_{jkl}^{BCD} = \tau_{kl}^{CD} = 0 .$$

We can conclude from the sufficient marginals that $\tau_{kl}^{CD} = 0$ because the sufficient marginal CD is missing and the marginal $x_{..kl}$ can not be derived from the only 3-factor marginal $x_{ijk.}$ included. A simpler way to express this is that the letter combination CD is not a subset of ABC. On the other hand τ_{ij}^{AB} is not necessarily 0, since the letter combination AB is a subset of ABC.

It is impossible to list all models. But it is possible to list the main types of hierarchical models. This is done in Table 4.1. This table also contains several characterizations of the models, which will be introduced later.

TABLE 4.1. All main types of hierarchical models for a 4-way contingency table and their interpretations.

Classification	Sequence of hierarchical models	Sufficient marginals	Interpretation	Type
	✓	ABC,ABD,ACD,BCD	-	S
	✓	ABC,ABD,ACD	-	S
	✓	ABC,ABD,CD	-	S
G, D		ABC,ABD	C⊥D|A,B	I
	✓	ABC,AD,BD,CD	-	S
		ABC,AD,BD	C⊥D|A,B	I
G, D		ABC,AD	D⊥B,C|A	II
G, D		ABC,D	D⊥A,B,C	III
G, D		ABC	D⊥A,B,C & D=u	XI
	✓	AB,AC,AD,BC,BD,CD	-	S
	✓	AB,AC,AD,BC,BD	C⊥D|A,B	I
	✓	AB,AC,AD,BC	D⊥B,C|A	II
G		AB,BC,CD,AD	A⊥C|B,D & B⊥D|A,C	IV
		AB,AC,BC,D	D⊥A,B,C	III
		AB,AC,BC	D⊥A,B,C & D=u	XI
	✓	AB,AC,AD	B⊥C⊥D|A	VII
G, D		AB,BC,CD	D⊥A,B|C & A⊥C,D|B	V
G, D	✓	AB,AC,D	D⊥A,B,C & B⊥C|A	VI
G, D		AB,AC	D⊥A,B,C & B⊥C|A & D=u	X
G, D		AB,CD	A,B⊥C,D	VIII
G, D	✓	AB,C,D	A,B⊥C,D & C⊥D	IX
G, D		AB,C	A,B⊥C,D & C⊥D & D=u	XII
G, D		AB	A,B⊥C,D & C⊥D & C=D=u	XIII
G, D	✓	A,B,C,D	A⊥B⊥C⊥D	XIV
G, D	✓	A,B,C	A⊥B⊥C⊥D & D=u	XV
G, D	✓	A,B	A⊥B⊥C⊥D & C=D=u	XVI
G, D	✓	A	A⊥B⊥C⊥D & B=C=D=u	XVII
G, D	✓	-	A⊥B⊥C⊥D & A=B=C=D=u	XVIII

A sequence of models is **hierarchically ordered** if all interactions, which are 0 in one model are also 0 in all subsequent models. The models marked by a ✓ in Table 4.1 thus form a sequence of hierarchically ordered models. Obviously we obtain a sequence of hierarchically ordered hypotheses by successively setting interactions and main effects to 0. The models marked with ✓ in Table 4.1 thus correspond to setting interactions and main effects equal to zero in the following order:

(1): $\tau_{ijkl}^{ABCD} = 0$ (2): $\tau_{jkl}^{BCD} = 0$ (3): $\tau_{ikl}^{ACD} = 0$

(4): $\tau_{ijl}^{ABD} = 0$ (5): $\tau_{ijk}^{ABC} = 0$ (6): $\tau_{kl}^{CD} = 0$

(7): $\tau_{jl}^{BD} = 0$ (8): $\tau_{jk}^{BC} = 0$ (9): $\tau_{il}^{AD} = 0$

(10): $\tau_{ik}^{AC} = 0$ (11): $\tau_{ij}^{AB} = 0$ (12): $\tau_{l}^{D} = 0$

(13): $\tau_{k}^{C} = 0$ (14): $\tau_{j}^{B} = 0$ (15): $\tau_{i}^{A} = 0$

Most - but not all - the models in Table 4.1 can be interpreted by statements describing independence, conditional independence, or uniform distribution over categories. The easiest way to obtain the interpretation of a given model is to draw the **association diagram** for the model. An association diagram is drawn by plotting the four variables A, B, C and D as dots, and then connecting any two dots if the corresponding variables are both represented in the same interaction. Thus A and B are connected if either ABCD, ABC, ABD or AB are among the sufficient marginals defining the model. The generic types af association diagrams are shown in fig. 4.1 together with the interpretation of the model.

In Table 4.1 is marked in the last column which of the types in Figure 4.1 the model belongs to. Note here, that several models can have the same association diagram and hence the same interpretation. Note also, that some of the models are equivalent to the saturated model. As an example of three models with the same association diagram and the same interpretation, consider the models with sufficient marginals

M_1: ABC, ABD , M_2: ABC, AB, AD, BD and M_3: AB, AC, AD, BC, BD

By drawing the association diagram for each of these three models and comparing with fig. 4.1, we can see that they are all of type I and therefore equivalent in the sense of their interpretation. From the association diagram it is easy to derive the common interpretation for M_1, M_2 and M_3. The rule is, as we recall, that two variables are independent, if they are not connected at all in the diagram, and conditionally independent, given a set of other variables, if they are only connected by a route passing through these variables. The rule of thumb is here to cover the conditioning variables by (literally) your thumb (or any other finger or a pencil). If you now see no connection between two variables, they are independent given the covered set of variables. By using this rule, we conclude, that the interpretation of type I is

$$C \perp D | A,B .$$

In the association diagram for the models ABC, AD and AB, AC, AD, BC, having the same diagram, namely type II, we can only reach D from B or C by a route passing A. Hence the interpretation is

$$D \perp B,C | A .$$

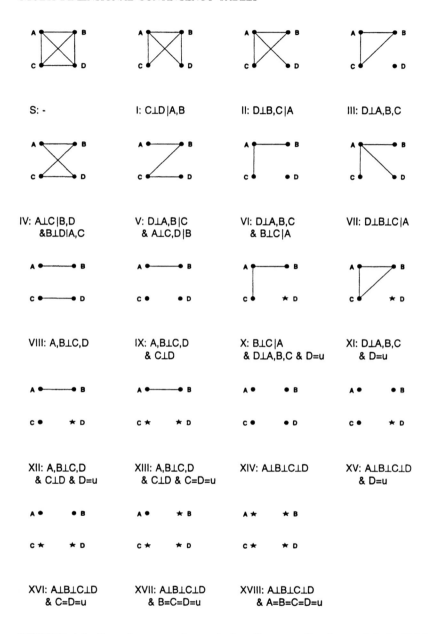

FIGURE 4.1. Generic types of association diagrams for the models in Table 4.1

It is important to note that two association diagrams can be of the same type, although they look different at first glance. An association diagram may thus sometimes seem to be missing in fig. 4.1. One example concerns the two models with sufficient marginals AB, AD, BC, CD and AB, AC, BD, CD. They are of the same type, namely type IV, as is readily seen by ex-changing the letters C and D.

As shown in fig. 4.2 the diagram is an "hour glass" (the one shown in fig 4.1) before the exchange of C and D and a "square" after.

FIGURE 4.2. Association diagrams for the models AB, AD, BC, CD and AB, AC, BD, CD.

Another example is the diagram, type II, for model ABC, AD, which becomes an "axe" if we exchange B and A, while in fig. 4.1 it is an "arrow". Finally the "Z" for model AB, BC, CD in fig 4.1 becomes a "three-winged castle" if C and D are interchanged.

For all models M with the same interpretation there is a most simple and a most complex model. There are thus a model M_{max} and a model M_{min}, such that $M_{min} \subset M \subset M_{max}$.

If a model M is the maximal model M_{max}, the model is called **graphic**. The minimal model M_{min} is defined by having all 2-factor interactions, corresponding to lines in the association diagram, as its sufficient marginals.

In order to define M_{max} we need the notion of a **clique** in an association diagram. If we visualize the association diagram as the description of 4 persons, A, B, C and D on a hiking trip, a line between two hikers mean that they communicate frequently during the trip. A clique is then a group of hikers, who all communicate, and is not a subgroup of a larger clique. All hikers in a clique thus communicate, and for all hikers outside the clique, there is at least one in the clique with whom they do not communicate. In exact terms a clique is a maximal subset of points in the association diagram, which are all connected. For type I in fig. 4.1, there are thus two cliques ABC and ABD, while for type II the cliques are ABC and AD.

A model is **graphic** if all interactions corresponding to cliques are included as sufficient marginals in the definition of the model. For the models M_1: ABC, ABD, M_2: ABC, AD, BD and M_3: AB, AC, AD, BC, BD with the same association diagram, M_1 is by this rule the graphical model. For models M_4: ABC, AD and M_5: AB, AC, AD, BC , also with the same association diagram, the graphical model is M_4. For the models in fig. 4.2 the minimal and the graphical model coincide because there are no 3-factor interactions and all included 2-factor interactions are cliques. The graphical models are marked by a "G" in Table 4.1, column 1.

The graphical models are equivalent with their interpretation since the association diagram can be derived from the interpretation. On the association diagram the cliques can then be identified and the model determined. On the other hand the interpretation can be derived from the association diagram, which is uniquely determined by the cliques, and thus the graphical model.

An important subset of the graphical models is the subset of **decomposable models**. The exact definition of a decomposable model is quite complicated. For our purpose it suffices to say that a model is decomposable if it is graphical and it does not contain any 4-point cycle anywhere in the association diagram, independently of how it is drawn. A 4-point cycle is a set of four points, which does not contain any triangles. For 4-way tables only the type IV models in fig. 4.2 and models obtained from these by exchange of letters, are graphical but not decomposable. For contingency tables of higher order, there are many graphical models which are not decomposable. In Table 4.1 the decomposable models are denoted by a "D" in column 1.

The main property, which holds for decomposable models and not for any other type of model, is that there are explicit expressions for the solutions of the likelihood equations, i.e. the estimated expected values can be expressed explicitly in terms of the sufficient marginals.

For all the decomposable model types in Table 4.1, the explicit expressions for the estimated mean values are shown in Table 4.2.

TABLE 4.2. Explicit expressions for the estimated expected values $\hat{\mu}_{ijkl}$ for all decomposable models in Table 4.1.

Sufficient marginals	Explicit expressions for $\hat{\mu}_{ijkl}$
ABC, ABD	$x_{ijk.} \cdot x_{ij.l} / x_{ij..}$
ABC, AD	$x_{ijk.} \cdot x_{i..l} / x_{i...}$
ABC, D	$x_{ijk.} \cdot x_{...l} / n$
ABC	$x_{ijk.} / L$
AB, AC, AD	$x_{ij..} \cdot x_{i.k.} \cdot x_{i..l} / x_{i...}^2$
AB, BC, CD	$x_{ij..} \cdot x_{.jk.} \cdot x_{..kl} / (x_{.j..} \cdot x_{..k.})$
AB, AC, D	$x_{ij..} \cdot x_{i.k.} \cdot x_{...l} / (n \cdot x_{i...})$
AB, AC	$x_{ij..} \cdot x_{i.k.} / (L \cdot x_{i...})$
AB, CD	$x_{ij..} \cdot x_{..kl} / n$
AB, C, D	$x_{ij..} \cdot x_{..k.} \cdot x_{...l} / n^2$
AB, C	$x_{ij..} \cdot x_{..k.} / (nL)$
AB	$x_{ij..} / (KL)$
A, B, C, D	$x_{i...} \cdot x_{.j..} \cdot x_{..k.} \cdot x_{...l} / n^3$
A, B, C	$x_{i...} \cdot x_{.j..} \cdot x_{..k.} / (n^2 L)$
A, B	$x_{i...} \cdot x_{.j..} / (nKL)$
A	$x_{i...} / (JKL)$

For nondecomposable models it is necessary to apply the iterative proportional fitting method to obtain the estimated expected values. The iterative proportional fitting method on the other hand, as is evident from Table 4.3 and the description of the method in section 3.3 for 3-way tables, provide the solutions after just one iteration in case there are explicit solutions. For computer programming, it is therefore easier to use the iterative proportional fitting method in all cases, rather than building a library of exact solutions for the decomposable models.

EXAMPLE 4.1. *Truck collisions.*
The data for this example were collected in England in two periods: The first from November 1969 to October 1971, the second from November 1971 to October 1973. The observations were the number of collisions involving trucks in the two periods. In addition to the period, which we shall call variable D with two categories, the collisions were also classified according to three more categorical variables:

  A: *Light conditions, with 3 categories Daylight, Night but illuminated road and Dark road.*

B: Movement, with two categories Parked and Moving.

C: Point of collision, with two categories in the Back or in the Front. (The front included the rare collisions in the sides of the trucks.)

Table 4.3 shows the observed number of collisions cross-classified over all four variables.

TABLE 4.3. The observed number of collisions cross-classified over Light condition, Movement, Point of collision and Period.

A: Light Conditions	B: Movement	C: Point of collision	D: Period Nov. 69 to Oct. 71	Nov. 71 to Oct. 73
Daylight	Parked	Back	712	613
		Front	192	179
	Moving	Back	2557	2373
		Front	10749	9768
Night, illuminated street	Parked	Back	634	411
		Front	95	55
	Moving	Back	325	283
		Front	1256	987
Night, dark street	Parked	Back	345	179
		Front	46	39
	Moving	Back	579	494
		Front	1018	885

Source: Leaflet from Transport and Road Research Laboratory. Department of Environment. Crowthorne. Berkshire. UK. October 1976.

The study was undertaken because a new safety measure for trucks was introduced in October 1971 and the traffic authorities wanted to know if this safety measure had any effect on the number of collisions and on where the trucks were hit during a collision. To answer one of these questions a model with conditional independence between the variables Period and Point of collision given the two variables, describing different environment conditions, would be an interesting model. This type of independence is conditional independence of variables C and D given A and B corresponding to the type I graphical model ABC, ABD.

Note: This is a typical situation where the correct model would be that the cell counts were independent Poisson distributed random variables. But, as we have seen, when we condition on the total number of collisions in both periods, we get the multinomial model (4.1).

The estimated expected numbers under the model ABC, ABD are shown in Table 4.4.

TABLE 4.4. The estimated expected numbers under the model ABC, ABD for the data in Table 4.3.

			D: Period	
A: Light Conditions	B: Movement	C: Point of collision	Nov. 69 to Oct. 71	Nov. 71 to Oct. 73
Day light	Parked	Back	706.2	618.7
		Front	197.8	173.3
	Moving	Back	2577.9	2352.1
		Front	10728.1	9788.9
Night, illuminated street	Parked	Back	637.5	407.5
		Front	91.5	58.5
	Moving	Back	337.2	270.8
		Front	1243.8	999.2
Night, dark street	Parked	Back	336.4	187.6
		Front	54.6	30.4
	Moving	Back	575.8	497.2
		Front	1021.2	881.8

The values in Table 4.4 are copied from a computer output, but can, in fact by the exact expressions in Table 4.2, be calculated directly from the marginals over C, over D and over both C and D. For example $x_{111\cdot} = 1325$, $x_{11\cdot 1} = 904$ and $x_{11\cdot\cdot} = 1696$ yields

$$\hat{\mu}_{1111} = \frac{1325 \cdot 904}{1696} = 706.2 \ .$$

The fit of the model is very satisfactory, as the closeness of the observed counts and the expected numbers under the model strongly suggests. The Z-test statistic for the model has observed value

$$z = 6.86, \quad df = 6 ,$$

The level of significance of this result is p=0.334, which is not significant at any reasonable level.

4.3 Choice of model

Altogether there are 28 hierarchical model types in Table 4.1. If we also count all those derived from the generic types in Table 4.1 by an interchange of letters, there are 166 different hierarchical models. Some of these are variations including uniform distribution over categories. Even if we exclude these, there remains 113 different hierarchical models to search among. It follows that it is of vital importance for a successful statistical analysis, that a sound procedure for the selection of a satisfactory model is employed.

In some situations we know beforehand, as was the case in Example 4.1, which model, or which small collection of models, is of interest. This is, for example, the case if there is a particular hypothesis, specified in terms of interactions being 0, we want to test. In this case we can use the test statistic (4.3), which for a 4-dimensional table has observed value

$$z(H) = 2 \sum_{i=1}^{I} \sum_{j=1}^{J} \sum_{k=1}^{K} \sum_{l=1}^{L} x_{ijkl} \left[\ln x_{ijkl} - \ln \hat{\mu}_{ijkl} \right], \quad (4.16)$$

We reject the hypothesis, if the level of significance

$$p = P(Q \geq z(H)), \quad (4.17)$$

is suitable small, where $Q \sim \chi^2(df(H))$ and $df(H)$ are the degrees of freedom for the approximating χ^2-distribution. If p is larger than a certain value, we accept H. Often H is accepted if p is larger than 0.05. In other situations there are two competing hypotheses H and H^* we want to compare. If this is the case, the purpose of the statistical analysis is to compare the fit of the data to the two models corresponding to the two hypotheses in question. Usually the task is to determine if a simpler model gives as good a fit as a more complex model. In this situation H and H^* must obey the relationship

$$H \subset H^*.$$

We can then use the sequential test statistic (4.5), which for a 4-dimensional table has observed value

$$z(H | H^*) = z(H) - z(H^*) = 2 \sum_i \sum_j \sum_k \sum_l x_{ijkl} \left[\ln \tilde{\mu}_{ijkl} - \ln \hat{\mu}_{ijkl} \right], \quad (4.18)$$

where $\tilde{\mu}_{ijkl}$ are the estimated expected numbers under H^*, and $\hat{\mu}_{ijkl}$ the estimated expected numbers under H. We would prefer the simpler model under H over the model under H^* if

$$p = P(Q \geq z(H|H^*)) \quad (4.19)$$

has a value larger than a specified level, for example p = 0.05, where

$$Q \sim \chi^2(df(H) - df(H^*)) \;,$$

since this indicates that the fit under H is as good as the fit under H^*.

In most situations we must, however, start by evaluating the fit of various models by means of the test statistic (4.16). But (4.16) can be combined with the sequential test (4.18) if the possible candidates for a fit of the data to the model can be listed in a manageable short list.

For models, which include interactions between 3 or more variables, it is in general difficult to interpret these interactions. Hence it is often a sensible first procedure to start with the model

$$AB, AC, AD, BC, BD, CD \;, \qquad (4.20)$$

which only contains 2-factor interactions and then by sequential tests try to omit one or more of these two-factor interactions. We can then use the test statistic (4.18), with level of significance (4.19), to evaluate if an interaction can be omitted. This method of course is based on acceptance of the model (4.20). If the model (4.20) can not be accepted, we have to go back to a more complicated model, for example the saturated, and then use sequential tests (4.18) or direct tests (4.16) to see if the 4-factor interaction and subsequently one or more of the three-factor interactions can be omitted without significantly worsening the fit of the model to the data. In this connection it is worth noting from Table 4.1, that relatively few model types with 3-factor interactions have interpretations in terms of independence or conditional independence. If the simplest model, which fits the data, does not have such an interpretation, and in addition contains interactions of dimension 3 or higher, very little can be said as a conclusion of the statistical analysis. There are, however, models with 3-factor interactions, which are of interest and where a useful statistical conclusion can be stated. Such an example is the model in Example 4.1 with sufficient marginals ABC, ABD, for which the interpretation is that variables C and D are independent given variables A and B.

EXAMPLE 4.1 (continued). *The model ABC, ABD fits the truck collision data in Table 4.34 well. In addition we show below in Table 4.56, by applying sequential tests, that no simpler, interesting model fits the data. Hence the conclusion is that the number of collisions in the back as compared to the number in the front of the truck, given the light and movement conditions, did not change from period 1 to period 2, i.e. before and after the introduction of the safety measure. The association diagram describing this independence is shown as Figure 4.3.*

MULTI-DIMENSIONAL CONTINGENCY TABLES

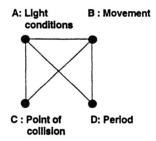

FIGURE 4.3 Association diagram for the truck collision data under the model ABC, ABD.

Table 4.5 shows the appropriate sequence of sequential test leading to the choice of the model ABC, ABD.

TABLE 4.5 A sequence of sequential tests for the data in Table 4.3

Model	Minus	z(H)	df	Level sign.	z(H|H*)	df	Level sign.
ABC, ABD, ACD, BCD	-	2.42	2	0.298	-	-	-
ABC, ABD, ACD	BCD	4.26	3	0.235	1.84	1	0.176
ABC, ABD, BCD ✓	ACD	5.19	4	0.269	2.76	2	0.251
ABC, ACD, BCD	ABD	14.92	4	0.005	12.49	2	0.002
ABD, ACD, BCD	ABC	37.78	4	0.000	35.36	2	0.000
ABC, ABD, CD ✓	BCD	6.60	5	0.252	1.41	1	0.235
ABC, BCD, AD	ABD	19.88	6	0.003	14.69	2	0.001
ABD, BCD, AC	ABC	39.78	6	0.000	34.60	2	0.000
ABC, ABD ✓	CD	6.85	6	0.335	0.26	1	0.612
ABC, AD, BD, CD	ABD	22.33	7	0.002	15.73	2	0.000
ABD, AC, BC, CD	ABC	41.49	7	0.000	34.90	2	0.000
ABC, AD, BD ✓	ABD	22.64	8	0.004	15.78	2	0.000
ABD, AC, BC	ACD	41.74	8	0.000	34.89	2	0.000
ABC, AD	BD	41.23	9	0.000	18.60	1	0.000
ABC, BD	AD	47.81	10	0.000	25.18	2	0.000
AB, AC, AD, BC, BD	ABC	57.57	10	0.000	34.93	2	0.000

With ✓, we have marked which model has been chosen in each step of the elimination procedure. It is worth noticing, that it is necessary to omit the two-factor interaction between variables C and D, before the association diagram becomes the picture in fig. 4.3. We may also note, that a procedure, where we try to omit *all 3-factor interactions* before starting to omit 2-factor interactions will not lead to the model ABC, ABD. According to this procedure, we would try to fit the models ABC, AD, BD, CD and BCD, AB, AC, CD first, both of which are clearly rejected. The procedure would then have stopped with model ABC, ABD, CD of type S having the same association diagram as the saturated model.

EXAMPLE 4.2. *Smoking habits and headaches.*
This data set is from an investigation carried out by the Danish Institute for Building Research in January 1983. A random sample of Danes was interviewed with respect to a number of issues. Among the categorical variables, which was the result of questions asked, consider.

A: *Smoking habits, with the categories Smoker and Non-smoker.*

B: *Age with categories Over 40 and Below 40.*

C: *Sex, with categories Men and Women.*

D: *Frequency of head aches, with categories More than once a week and Less the once a week.*

Table 4.6 shows the 2×2×2×2 contingency table of observed numbers for each combination of the categories of the four variables.

TABLE 4.6. A random sample of Danes cross-classified according to the variables: Smoking habits, Age, Sex and Frequency of head aches.

A: Smoking habits	B: Age	C: Sex	D: Frequencies of head aches	
			More than once a week	Less than once a week
Smoker	Under 40	Men	11	142
		Women	45	83
	Over 40	Men	11	145
		Women	15	76
Non-smoker	Under 40	Men	8	117
		Women	29	89
	Over 40	Men	7	113
		Women	8	80

Source: Unpublished data from the Danish Institute for Building Research.

For this data set we shall use a search strategy where we start by testing the fit of the model AB, AC, AD, BC, BD, CD with all 2-factor interactions included, but without interactions of higher dimensions and successively try to omit 2-factor interactions from the model.

The test statistic (4.16) for the model AB, AC, AD, BC, BD, CD has observed value

$$z(H) = 6.63 , df = 5$$

which corresponds to the level of significance

$$p = P(Q \geq 6.63) = 0.249 .$$

Hence we can accept that a model without 3-factor and 4-factor interactions describes the data.

The procedure is now to try to omit two-factor interactions one by one. The order in which we omit 2-factor interactions is determined by the level of significance for each new model. We could also have used the level of significance for the sequential tests, comparing each model with and without a 2-factor interaction included. That the interaction between A and B is the first we try to omit, is, therefore, due to the fact, that the level of significance is 0.344 for the model AC,AD,BC,BD,CD - which is the one, where AB is omitted - and that this level of significance is higher than for the other five models with two-factor interactions excluded. The test statistics and levels of significance for the steps of this search procedure are shown in Table 4.7.

TABLE 4.7. Test statistics and levels of significance for the models obtained by successively omitting 2-factor interactions from the model AB, AC, AD, BC, BD, CD.

Model	Minus	z(H)	df	Level sign.	$z(H\|H^*)$	df	Level sign.
AB, AC, AD, BC, BD, CD		6.63	5	0.249	-	-	-
AC, AD, BC, BD, CD	AB	6.76	6	0.344	0.13	1	0.723
AC, AD, BD, CD	BC	9.08	7	0.247	2.32	1	0.128
AD, BD, CD	AC	12.30	8	0.138	3.23	1	0.072
BD, CD, A	AD	15.63	9	0.075	3.33	1	0.068
CD, A, B	BD	31.82	10	0.000	16.19	1	0.000
A, B, C, D	CD	85.25	11	0.000	53.43	1	0.000

For this data set we are fortunate, that in order to reach a conclusion, it is not a question of choosing the level of significance for the various tests very carefully, for example in the interval between 0.05 and 0.01. From Table 4.7 it is quite obvious, that it is the step from model BD, CD, A to model CD, A, B, which seriously makes the fit of the model worse. The model BD, CD, A is thus the most simple model, which gives us a satisfactory fit. (A uniform distribution over the *categories of A, i.e equally many smokers and non-smokers marginally, is of no interest in this connection.) The association diagram for the model with sufficient marginals BD, CD and A is shown in Figure 4.4.*

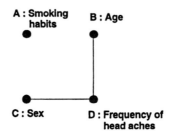

FIGURE 4.4. Association diagram for the model BD, CD, A

The association diagram shows that the interpretation is

$$A \perp B,C,D \quad \& \quad C \perp B|D.$$

Smoking habits are thus, it seems, independent of age as well as sex. Most important is, of course, that there is no connection between smoking habits and frequency of headaches, if these data are to be trusted. The conditional independence of B and C given A is not very interesting. We must at least not fall for the "causality trap" of saying that frequency of headaches "explains" that age and sex are independent variables. This last independence is just a consequence of the simple fact that there are about the same number of men and women among those under 40, and the same is true for those over 40. It is more important to note, that frequency of headaches depends on both age and sex. This is well illustrated by the estimated 2-factor interactions shown in Table 4.8.

TABLE 4.8. Estimated 2-factor interactions between variables B and D and between variables C and D.

$\hat{\tau}^{BD}_{jl}$		l = 1	2
		Frequent headaches	Infrequent headaches
j=1	Under 40	0.195	-0.195
2	Over 40	-0.195	0.195

$\hat{\tau}^{CD}_{kl}$		l = 1	2
		Frequent headaches	Infrequent headaches
k=1	Men	-0.355	0.355
2	Women	0.355	-0.355

Relatively fewer men thus suffer from frequent headaches, and the frequency of headaches declines over the years.

4.4 Diagnostics

In situations where the model search and the hypotheses tested does not lead to a satisfactory simple model, it is important to have diagnostics to tell us which cells, or which combinations of categories, we should inspect closer. The most important of these diagnostics are the **standardized residuals**, defined as

$$r_{ijkl} = \frac{x_{ijkl} - \hat{\mu}_{ijkl}}{\hat{\sigma}_{ijkl}}, \quad (4.21)$$

where

$$\hat{\sigma}^2_{ijkl} = \hat{\text{var}}[X_{ijkl} - \hat{\mu}_{ijkl}] \quad (4.22)$$

and the $\hat{\mu}_{ijkl}$ are the estimated expected numbers under the model. The symbol vâr means that the estimated values of the parameters are inserted in the variance formula given below as (4.23). The distribution of r_{ijkl} is approximately a standard normal distribution, so values higher than 2 numerically are indications of a poor model fit. The formula for (4.22) is known, although a bit complicated, namely

$$\sigma^2_{ijkl} = \text{var}[X_{ijkl} - \hat{\mu}_{ijkl}] = \mu_{ijkl}(1 - h_{ijkl}), \quad (4.23)$$

where the h_{ijkl}'s are the diagonal elements of the "hat"-matrix **H**, defined by the general formula (2.45), with the necessary adjustments to the log-linear model for a 4-way table. For the present we shall not go into more details, as regards the exact form of **H**. The necessary calculations must at any rate be done by a computer, and, as noted in section 2.9, the formulas are well suited for computer programming.

Sometimes it is also helpful to look at the **standardized interactions**, i.e the ML-estimates of the interaction parameters divided by their standard error. For the 3-factor interaction between variables A, B and C, for example, the standardized interaction is defined as

$$\hat{\omega}^{ABC}_{ijk} = \frac{\hat{t}^{ABC}_{ijk}}{\hat{\sigma}^{ABC}_{ijk}} \quad (4.24)$$

where

$$\left(\hat{\sigma}^{ABC}_{ijk}\right)^2 = \text{vâr}[\hat{t}^{ABC}_{ijk}].$$

By the general theory in chapter 2, ML-estimates are approximately normally distributed. Hence the quantities (4.24) are approximately normally distributed with mean 0 and variance 1. So again values numerically larger than 2 call for a closer inspection. It follows that if the value of (4.24) is larger than 2 or smaller than -2 for values i, j and k there is a positive or negative connection between the variables A, B, and C, which manifests itself especially for cells where the combination of the subscripts contain i, j and k, i.e. cells (ijk1), (ijk2), ... , (ijkL).

Finally there is **Cook's distance**, which is a measure for how much a cell contributes to the values of the ML-estimates for the log-linear parameters. We return to Cook's distance in more detail in chapter 7. For the present, it suffices to say the following: Let τ be the vector of log-linear parameters and M the matrix defined by formula (2.43). Then Cook's distance is given by

$$C_{ijkl} = \frac{1}{N}\left(\hat{\tau}-\hat{\tau}^{(ijkl)}\right)' n M\left(\hat{\tau}-\hat{\tau}^{(ijkl)}\right) \qquad (4.25)$$

where N is the number of unconstrained τ's and $\hat{\tau}^{(ijkl)}$ is the vector of log-linear parameters estimated without using the data in cell (ijkl). Since the variance matrix of $\hat{\tau}$ is M^{-1}/n, Cook's distance is a scaled difference between the estimates we get if all cells are included in the estimation and the estimates we get if all cells except cell (ijkl) are included. It follows that Cook's distance measures the **influence** of the data in cell (ijkl) on the estimated parameter values. It has been shown (cf. bibliographical notes) that an approximation to Cook's distance is the following

$$C_{ijkl} = \frac{1}{N}r_{ijkl}^2 \frac{\hat{h}_{ijkl}}{1-\hat{h}_{ijkl}} . \qquad (4.26)$$

Cook's distance is thus large if a residual is large, but also if the value of $\hat{h}_{ijkl}$ is close to 1. The quantity $\hat{h}_{ijkl}$ is called the **leverage**, and is an important indicator for any cell of the contingency table. It follows from (4.23), for example, that the reduction in the variance of the residual as compared to the "binomial" variance $n \pi (1 - \pi)$ is large if the leverage is close to 1 and small if the leverage is close to 0. From (4.26) it further follows that the influence of a cell on the parameter estimates can be large, even though the residual is moderate, namely if the leverage is close to 1.

Example 4.3. *Employment status.*
In 1974 the Danish National Institute of Social Research made an investigation of 1314 employee's who had left their job during the second half of the year. These lay-offs we cross-classified according to the following three categorical variables:

  A: *Employment status on January 1, 1975, with categories Got a new job and Still unemployed.*

  B: *Cause of lay-off, with categories Closure of firm (incl. reduction of the number of employees) and Replacement by another employee.*

C: *The length of the employment prior to the lay-off, with 5 categories.*

The counts obtained by cross-classifying these 3 variables are shown in Table 4.9.

TABLE 4.9. A sample of employees cross-classified according to Employment status on January 1st, Cause of lay-off and Length of employment prior to the lay-off.

		C: Employment status on January 1, 1975	
A: Length of employment	B: Cause of lay-off	Got a new job	Still unemployed
Less then 1 month	Closure	8	10
	Replacement	40	24
1-3 month	Closure	35	42
	Replacement	85	42
3 month to 1 year	Closure	70	86
	Replacement	181	41
1-2 years	Closure	62	80
	Replacement	85	16
2-5 years	Closure	56	67
	Replacement	118	27
More than 5 years	Closure	38	35
	Replacement	56	10

Source: Kjær, A. (1978): Redundancy in the Labour Market. Literature and concepts. (In Danish). Study no. 36. The Danish National Institute of Social Research. Copenhagen: Teknisk Forlag.

TABLE 4.10. Test statistics for different models applied to the data in Table 4.9.

Model	Minus	z(H)	df	Level sign.	z(H|H*)	df	Level sign.
AB, AC, BC		9.02	5	0.108	-	-	-
AB, BC	AC	24.63	10	0.006	15.62	5	0.008
AC, BC	AB	64.62	6	0.000	55.61	1	0.000
AB, AC	BC	165.92	10	0.000	156.91	5	0.000

Table 4.10 shows all the test statistics, if we try to omit one of the 2-factor interactions from the model. At first glance there is no indication of a reduction from the model AB, AC, BD, with an association diagram where all the variables are connected. If we take a look at the standardized residuals in Table 4.11, we gain some hope, however. The residuals are computed under the model AB, BC,

because we are primarily interested in a connection between the chance of getting a job relatively soon again, and the length of employment prior to the lay-off.

TABLE 4.11. Standardized residuals for the employment data in Table 4.9 under the model AB, BC.

		C: Employment status on January 1, 1975	
A: Length of employment	B: Cause of lay-off	Got a new job	Still unemployed
Less then 1 month	Closure	-0.11	+0.11
	Replacement	-3.12	+3.12
1-3 month	Closure	-0.04	+0.04
	Replacement	-3.29	+3.29
3 month to 1 year	Closure	-0.23	+0.23
	Replacement	+1.55	-1.55
1-2 years	Closure	-0.55	+0.55
	Replacement	+1.63	-1.63
2-5 years	Closure	-0.04	+0.04
	Replacement	+1.12	-1.12
More than 5 years	Closure	+1.17	-1.17
	Replacement	+1.42	-1.42

The residuals in Table 4.11 shows that there are two model modifications which seems to point to a satisfactory model. One is to study only individuals with variable B at level 1, i.e. those who have been laid-off due to closures and reductions. If this is the reason for the lay-off all the residuals are insignificant in value, and we should expect independence between variables A and C. We may write this as $A \perp C \mid B(1)$, B(1) meaning B at level 1.

The second possible model modification is based on the observation that the residuals are all of the same size and magnitude for variable A at levels 1 and 2, and all residuals are insignificant for variable A at levels 4 to 6. If we thus split the persons in two groups: Those with a very short employment prior to the lay-off and those with a reasonably long employment, i.e. over 3 month at the time of the lay-off, we should expect independence between variables A and C given B. That this is true is confirmed by Table 4.12, which shows the same test statistics as in Table 4.10, but now only for persons with length of employment over 3 months at time of lay-off.

TABLE 4.12. Test statistics for different models applied to the data in Table 4.10, when only persons with an employment length of more than 3 months, prior to the lay-off, are included in the data.

Model	Minus	z(H)	df	Level sign.	z(H|H*)	df	Level sign.
AB, AC, BC		0.52	3	0.914	-	-	-
AB, BC	AC	2.19	6	0.902	1.67	3	0.644
AC, BC	AB	18.46	6	0.005	17.94	3	0.001
AB, AC	BC	154.96	4	0.005	154.44	1	0.001

4.5 Model search strategies

For contingency tables of higher dimensions than 4, it is impossible to keep track of all hierarchical models. Even to decide in which sequence one should test the fit of the graphical models is a difficult problem. There are good reasons for limiting attention to the graphical models. They are for example the maximal models with a given interpretation. Hence further reductions within the class of hierarchical models with the same association diagram would not change the interpretations in terms of independence and conditional independence, but only in some cases reduce the number of higher order interactions, making the interpretation of the estimated log-linear parameters easier.

In the years 1983 to 1985 a number of search strategies among graphical models were introduced (see the bibliographical notes). In Example 4.5 we shall for a 5-way table demonstrate the following simple strategy. Consider first the association diagram for the saturated model, where all points are connected. From this base model the steps of the procedure are as follows.

Step 1. Try to remove all lines in the association diagram one by one. This gives rise to a number of new graphical models. The model among these with the highest level of significance, thus fitting the data best, is then taken as the next base model.

Step 2. From the base model chosen in step one (with one line missing in the association diagram) we then try to remove each of the remaining lines one by one again giving rise to a number of new graphical models. The model among these with the highest level of significance is then the next base model. It has of course an association diagram with two lines missing in the association diagram.

Steps 3 and the following: In each new step the procedure in step 2 is repeated until the next base model has a level of significance which clearly indicates, that the model fit is unsatisfactory. The model selected in the previous step is thenthe final model.

EXAMPLE 4.4. *Tax evasion data.*
The study was based on a sample of employed men in the age interval 18 to 67. The key question asked was whether they in the preceding 12 months had done any work, which before they would have paid a craftsman to do. This question - among others - should illustrate the amount of tax evasion in Denmark in the building industry, a phenomenon widely known as "black work". The responses "yes" and "no" to this question are recorded in the table below as variable D. Table 4.14 shows a 5-way contingency table formed by variable D and the variables:

A: *Type of residence, with categories Apartment and House.*

B: *Employment, with categories Skilled blue collar, Unskilled blue collar and White collar.*

C: *Mode of residence, with categories Renter and Owner.*

E: *Age group, with categories Under 30, 31-45 and 46-67.*

TABLE 4.13. A sample of Danish men cross-classified according to Type of residence, Employment, Mode of residence, Response to without craftsman and Age.

A: Type of residence	B: Employment	C: Mode of residence	D: Response to work without craftsman	E: Age interval		
				Under 30	31-45	46-67
Apartment	Skilled blue collar	Renter	Yes	18	15	6
			No	15	13	9
		Owner	Yes	5	3	1
			No	1	1	1
	Unskilled blue collar	Renter	Yes	17	10	15
			No	34	17	19
		Owner	Yes	2	0	3
			No	3	2	0
	White collar	Renter	Yes	30	23	21
			No	25	19	40
		Owner	Yes	8	5	1
			No	4	2	2
House	Skilled blue collar	Renter	Yes	34	10	2
			No	28	4	6
		Owner	Yes	56	56	35
			No	12	21	8
	Unskilled blue collar	Renter	Yes	29	3	7
			No	44	13	16
		Owner	Yes	23	52	49
			No	9	31	51
	White collar	Renter	Yes	22	13	11
			No	25	16	12
		Owner	Yes	54	191	102
			No	19	76	61

Source: Edwards and Kreiner (1983).

The results of the tests during step 1 of the procedure are shown in Table 4.14.

TABLE 4.14 Significance tests for removal of 2-factor interactions one by one from the saturated model for the data in Table 4.13.

Model	Interaction omitted	Z(H)	df	Level of significance
ACDE, BCDE	AB	32.37	24	0.118
ABDE, BCDE	AC	541.68	18	0.000
ABCE, BCDE	AD	19.81	18	0.344
ABCD, BCDE	AE	76.09	24	0.000
ACDE, ABDE	BC	41.06	24	0.016
ACDE, ABCE	BD	52.98	24	0.001
ABCD, ACDE	BE	106.36	32	0.000
ABCE, ABDE	CD	73.71	18	0.000
ABCD, ABDE	CE	194.32	24	0.000
ABCD, ABCE	DE	38.77	24	0.029

From Table 4.14 we must - according to the procedure - choose the model with sufficient marginals ABCE and BCDE. It has the association diagram shown in fig. 4.5 (a). Since we have removed the line between A and D, the interpretation of the model is $A \perp D \mid B,C,E$. The next step is with model ABCE, BCDE as base model to try to remove the remaining lines in Figure 4.5 (a) one by one. The 9 tests corresponding to these attempts to remove lines in the association diagram are shown in Table 4.15.

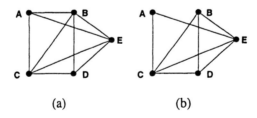

(a)   (b)

FIGURE 4.5. Association diagrams for models ABCE, BCDE and BCDE, ACE.

TABLE 4.15 Significance tests for removal of 2-factor interactions one by one from the model ABCE, BCDE.

Model	Interaction omitted	Z(H)	df	Level of significance
BCDE, ACE	AB	35.64	30	0.256
BCDE, ABE	AC	575.75	27	0.000
BCDE, ABC	AE	77.02	30	0.000
ACE, ACE, BDE, CDE	BC	51.85	36	0.040
ABCE, CDE	BD	55.25	30	0.003
ABC, ADE, BCD, CDE	BE	109.22	42	0.000
ABCE, BDE	CD	107.79	27	0.000
ABC, ABE, BCE, BCD	CE	204.19	36	0.000
ABCE, ABC	DE	39.71	30	0.111

We now conclude that the line between A and B can be removed, since the model BCDE, ACE without this line in the association diagram has the largest level of significance, namely $p = 0.256$. The association diagram for this model is shown as Figure 4.5 (b). The interpretation is now, that $A \perp B,D \mid C,E$. We still have a satisfactory fit by the new base model: BCDE, ACE, so we can go on with step 3 of the procedure.

From the new base model BCDE, ACE we again try to remove lines in the association diagram fig 4.5 (b). The required tests are shown in Table 4.16.

TABLE 4.16 Significance tests for removal of 2-factor interactions one by one from the model BCDE, ACE.

Model	Interaction omitted	Z(H)	df	Level of significance
BCDE, AE	AC	594.45	33	0.000
BCDE, AC	AE	88.06	34	0.000
ACE, BDE, CDE	BC	69.12	42	0.005
ACE, BCE, CDE	BD	70.07	42	0.004
ACE, BCD, CDE	BE	120.20	46	0.000
ACE, BCE, BDE	CD	122.61	39	0.000
BCE, BCD, AE, AC,	CE	235.32	44	0.000
ACE, BCD, BCE	DE	54.53	42	0.093

The new base model is now ACE, BCD, BCE with a level of significance $p = 0.093$. The association diagram for this model is shown as fig 4.6. There are two *interpretations for this model, B, D ⊥ A | C, E and D ⊥ A,E | B, C.*

For the model in Figure 4.6 there are three cliques ACE, BCD and BCE and they are all included in the model with their full 3-factor interactions, which is the requirement for a model being graphical. Also note that it is decomposable, as the association diagram does not exhibit any 4-cycle.

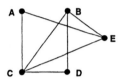

FIGURE 4.6. Association diagram for model ACE, BCD, BDE.

4.6 Bibliographical notes

As for chapter 3 the general references for log-linear models are Bishop, Fienberg and Holland (1975), Haberman (1978), (1979), Christensen (1990), Agresti (1990) and Andersen (1994). For graphical models the general references are Whittager (1990), Edwards (1995) and Lauritzen (1996). Goodmans fundamental contributions to the theory of log-linear models are collected in Goodman(1978).

The definition of hierarchical models and the classification of models as decomposable and non-decomposable was given by Goodman (1970) although he called the decomposable "elementary". He also showed that only the decomposable models have an explicit solution to the likelihood equations in terms of marginals. The name "decomposable" was suggested by Haberman (1974), see also Andersen (1974). The notion of a sufficient marginal is also due to Goodman (1970). These and other results were fully developed mathematically by Haberman (1974). Association diagrams were introduced in Goodman (1972), (1973). The theory for association diagrams was developed by Darroch, Lauritzen and Speed (1980), who introduced the graphical models and gave the conditions for a graphical model to be decomposable. Formulas for the standardized residuals was first given by Haberman (1974).

Model search strategies for multiple contingency tables has been studied by Edwards and Kreiner (1983) and by Edwards and Havranek (1985), (1987).

4.7 Exercises

[In some of these exercises it is necessary to apply statistical packages like SPSS, BMDP or SAS to compute test statistics and parameter estimates. For readers

MULTI-DIMENSIONAL CONTINGENCY TABLES

without access to such packages a number of key test statistics and parameter estimates are for selected exercises given in the Appendix]

4.1 In the two tables below, both extracted from the data base of the Danish Welfare Study, the variables Sex, Social group and Urbanization is compared with two typical Yes-No questions from the study: (1) Do you own a color-TV? and (1) Do you have a washing machine?

For each table:

(a) Try to find as simple a log-linear model as possible that fits the data.

(b) Draw the association diagram and interpret the model.

(1)

A: Owns a Color-TV	B: Sex	C: Urbanization	D: Social group			
			I-II	III	IV	V
Yes	Male	Copenhagen	12	23	36	19
		Cop: Suburbs	36	46	39	16
		Three big cities	28	25	49	23
		Other cities	75	112	119	90
		Countryside	24	90	63	80
	Female	Copenhagen	2	12	60	33
		Cop: Suburbs	9	20	76	43
		Three big cities	7	20	52	35
		Other cities	21	79	135	140
		Countryside	12	27	86	77
No	Male	Copenhagen	25	21	60	39
		Cop: Suburbs	46	23	47	34
		Three big cities	17	33	46	40
		Other cities	71	79	152	141
		Countryside	36	212	149	206
	Female	Copenhagen	12	30	76	43
		Cop: Suburbs	16	31	69	33
		Three big cities	8	26	68	54
		Other cities	21	70	154	188
		Countryside	16	60	150	200

(2)

A: Owns a washing machine	B: Sex	C: Urbanization	D: Social group			
			I-II	III	IV	V
Yes	Male	Copenhagen	9	13	15	9
		Cop: Suburbs	58	44	48	24
		Three big cities	38	38	51	26
		Other cities	118	142	168	108
		Countryside	56	241	173	213
	Female	Copenhagen	4	15	28	13
		Cop: Suburbs	19	32	76	31
		Three big cities	10	25	70	38
		Other cities	29	115	197	179
		Countryside	25	73	200	208
No	Male	Copenhagen	28	31	81	49
		Cop: Suburbs	24	25	38	26
		Three big cities	7	20	44	37
		Other cities	28	49	103	123
		Countryside	4	61	39	73
	Female	Copenhagen	10	27	108	63
		Cop: Suburbs	6	19	69	45
		Three big cities	5	21	50	51
		Other cities	13	34	92	149
		Countryside	3	14	36	69

4.2 The table below shows the Danish Welfare Study cross-classified according to

A: Alcohol consumption, with categories 0-1 units a day, 2-10 units a day and 10 or more units a day.
(A unit is roughly one 33 cl bottle of beer of strength 4% vol. or its equivalent.)

B: Social group with four categories (cf. Example 1.1).
C: Marriage status, with categories Married and Not married.
D: Age with categories: 20-39 years old and 40-69 years old.

(According to modern Danish thinking "married" includes "living in a permanent partnership".)

A: Alcohol consumption	B: Social group	C: Marriage status	D: Age 20-39	D: Age 40-69
0-1 units a day	I-II	Married	43	38
		Not married	11	6
	III	Married	112	158
		Not married	35	33
	IV	Married	195	169
		Not married	76	44
	V	Married	135	216
		Not married	72	61
2-10 units a day	I-II	Married	102	83
		Not married	29	21
	III	Married	147	166
		Not married	60	33
	IV	Married	242	182
		Not married	130	34
	V	Married	100	169
		Not married	83	33
More than 10 units a day	I-II	Married	47	51
		Not married	18	7
	III	Married	53	59
		Not married	20	12
	IV	Married	54	56
		Not married	65	16
	V	Married	31	43
		Not married	47	16

(a) Start both with the saturated model and with the model AB, AC, AD, BC, BD, CD as first model in your search for a simple model to describe the data. Give an account of your experiences with the two approaches.

The next table shows some of the standardized estimates for log-linear parameters in the model ACD, AB, BC, BD.

$\hat{\omega}^{ACD}_{ikl}$			l=1	2
i=1		k=1	+3.437	-3.437
		2	-3.437	+3.437
	2	k=1	-0.515	+0.515
		2	+0.515	-0.515
	3	k=1	-2.362	+2.362
		2	+2.362	-2.362

$\hat{\omega}^{AB}_{ij}$	j=1	2	3	4
i=1	-8.654	+1.363	+2.891	+8.322
2	+1.843	+0.022	+0.838	-3.034
3	+7.224	-1.146	-3.033	-4.201

(b) Use these values to describe the dependencies in the table.

In the table below the standardized residuals for the model ACD, BC, BD are shown.

A: Alcohol consumption	B: Social group	C: Marriage status	D: Age 20-39	40-69
0-1 units a day	I-II	Married	-4.497	-5.775
		Not married	-2.277	-0.129
	III	Married	-0.623	-0.655
		Not married	-0.398	+0.294
	IV	Married	+0.600	-0.015
		Not married	-0.730	+0.017
	V	Married	+3.699	+4.838
		Not married	+2.631	+1.000
2-10 units a day	I-II	Married	+2.325	+0.917
		Not married	-0.309	+3.734
	III	Married	+0.592	-0.343
		Not married	+0.400	+1.548
	IV	Married	+1.185	+0.837
		Not married	+0.617	-0.633
	V	Married	-3.901	-1.163
		Not married	-0.808	-2.930
More than 10 units a day	I-II	Married	+4.419	+5.368
		Not married	+0.889	+1.339
	III	Married	+1.558	+0.026
		Not married	-1.980	+0.268
	IV	Married	-2.969	-0.790
		Not married	+0.464	-0.139
	V	Married	-1.873	-3.153
		Not married	+0.636	-1.147

(c) Use this table to explain why the model ACD, BC, BD does not fit the data.

4.3 In 1983 the Danish Institute for Building Research investigated the indoor climate in Danish homes. A random sample of 1968 persons above the age of 16 were selected for interviewing. Here we consider the following variables:

A: Indoor temperature (normal for the house), with categories Under 23° and Over 23° (Celsius).
B: Age, with categories Under 40 and Over 40.
C: Moisture or mould at the walls, with categories Yes and No.
D: Irritation or dryness of the throat, with categories Yes and No.

The resulting 4-way contingency table was.

			D: Irritation of throat	
A: Temperature	B: Age	C: Moisture	Yes	No
Under 23°	Under 40	Yes	4	22
		No	24	607
	Over 40	Yes	3	12
		No	52	684
Over 23°	Under 40	Yes	3	6
		No	20	219
	Over 40	Yes	1	3
		No	34	274

(a) Use both direct and sequential test statistics to find an as simple as possible model which fits the data. Comment on the use of direct or of sequential test statistics in the model search procedure.

(b) Draw the association diagram for the chosen model and interpret the diagram.

The table below shows the standardized 2-factor interactions between variable D and the other three variables in the model AD, BD, CD.

$\hat{\omega}^{AD}$	l=1	2
i=1	-3.420	+3.420
2	+3.420	-3.420
$\hat{\omega}^{BD}$	l=1	2
j=1	-2.412	+2.412
2	+2.412	-2.412
$\hat{\omega}^{CD}$	l=1	2
k=1	+3.589	-3.589
2	-3.589	+3.589

(c) Use these values to describe the way the variables A, B and C interact with Irritation of the throat.

(d) Compare the sizes of the standardized estimates in (c) with the conclusions drawn in (a).

4.4 In a retrospective study of cancer in the ovary the Survival (variable B) by 10 years after being operated was recorded for 299 women together with the values of the variables:

A: Stage of the cancer at the time of operation, with categories Early and Advanced.
B: Operation mode, with categories Radical and Limited.
D: X-ray treatment applied, with categories No and Yes.

The observed numbers were:

A: Stage	B: Operation	C: Survival by 10 years	D: X-ray treatment No	D: X-ray treatment Yes
Early	Radical	No	10	17
		Yes	41	64
	Limited	No	1	3
		Yes	13	9
Advanced	Radical	No	38	64
		Yes	6	11
	Limited	No	3	13
		Yes	1	5

(a) Find a simple model that fits the data.

(b) Describe what the model tells you about the factors of importance for predicting the survival by 10 years after this type of operation.

(c) Compute the expected values under the model and comment on them.

(d) If

$$\lambda_{ijkl} = \ln\left(E[X_{ijkl}]\right),$$

show that under the model AC, B, D the 2-factor interactions between variables A and C are given by.

$$\tau_{ik}^{AC} = \bar{\lambda}_{i\cdot k\cdot} - \bar{\lambda}_{i\cdots} - \bar{\lambda}_{\cdot\cdot k\cdot} + \bar{\lambda}_{\cdots\cdot}$$

and use this result to estimate and interpret the 2-factor interactions between Survival and Stage.

4.5 From the report used in Example 3.1, one can also extract the following table, showing the connection between A: Sex, B: Age, C: Whether the family has a video recorder or not and D: Whether the teenager has been to a movie theater within the last month.

			D: Been to a movie	
A: Sex	B: Age	C: Has video	Yes	No
Boy	7-9	No	18	31
		Yes	49	68
	10-12	No	25	29
		Yes	56	57
	13-15	No	28	33
		Yes	63	45
Girl	7-9	No	15	35
		Yes	44	81
	10-12	No	21	28
		Yes	62	73
	13-15	No	28	18
		Yes	79	47

(a) Start by with the model AB, AC, AD, BC, BD, CD and use direct and sequential tests to select a simple model to describe the data.

(b) Interpret the model based on the association diagram.

(c) Use the formula in exercise 4.4 to estimate the 2-factor interactions (if any) in the selected model.

(d) Comment on any 2-factor interactions you would expect to be non-zero, but are in fact zero, in the selected model.

4.6 The Danish Council for Road Safety has collected the following table of number of traffic accidents in Jutland. The involved variables were: A: Time of the accident, B: Number of parts involved in the accident, C: Whether the driver was under influence of alcohol and D: The direction of the road. The last variable was of special interest because main roads (for example all 4-lane freeways), due to the geography of Jutland, in general go North-South while secondary roads go East-West.

A: Time	B: Number of parts	C: Under influence of alcohol	D: Direction N-S	D: Direction E-W
Morning	One	Yes	29	38
		No	6	4
	More than one	Yes	294	206
		No	7	5
Afternoon	One	Yes	13	14
		No	3	7
	More than one	Yes	120	102
		No	8	15
Evening	One	Yes	15	21
		No	11	16
	More than one	Yes	53	52
		No	18	12
Night	One	Yes	17	23
		No	26	20
	More than one	Yes	16	16
		No	4	11

(a) Start your search for a simple model which fits the data with the model AB, AC, AD, BC, BD, CD and comment on the levels of significance for the Z(H)'s and those for the Z(H|H*)'s which you evaluated during the search procedure.

(b) Consider the three models (1) AB, AC, BC, BD, (2) AB, AC, BC, D and (3) AB, AC, BC. Compare their association diagrams, their interpretations and their levels of significance. Draw your conclusions.

(c) How would you report your findings concerning differences between accidents on N-S roads and on E-W roads to Road Authorities?

4.7 The Danish Gallup Institute addressed the issue of corporale punishment of children in a survey in the late 70's. From this survey the following table can be extracted for the variables A: Having the opinion that you can punish your children or the opinion that you can not, B: Memory of being punish yourself as a child, C: Education and D: Age.

A: Punish as a child	B: Memory of punishment	C: Education	D: Age 15-24	25-39	40-
No	Yes	Elementary	1	3	20
		Secondary	2	8	4
		High	2	6	1
	No	Elementary	26	46	109
		Secondary	23	52	44
		High	26	24	13
Yes	Yes	Elementary	21	41	143
		Secondary	5	20	20
		High	1	4	8
	No	Elementary	93	119	324
		Secondary	45	84	56
		High	19	26	17

(a) Use direct or sequential tests to select a simple model to describe the data.

(b) Draw the association diagram for the selected model.

(c) Is the selected model graphical and/or decomposable?

(d) What are the interpretations of the selected model?

4.8 Consider the following five variables in the Danish Welfare Study: A: Sex, B: Age, C: Family taxable income, D: Employment sector and E: Whether there is a freezer in the household. The age categories were

    Old: Over 40.
    Young: Under 40.

The income intervals were:

    Low: Under 60 000 D.kr.
    Medium: Between 60 000 Dkr. and 100 000 Dkr.
    High: Over 100 000 D.kr.

MULTI-DIMENSIONAL CONTINGENCY TABLES

A: Sex	B: Age	C: Income	D: Sector	E: Freezer in the household Yes	No
Male	Old	High	Private	152	39
			Public	82	18
		Medium	Private	135	31
			Public	35	12
		Low	Private	89	45
			Public	20	9
	Young	High	Private	259	46
			Public	101	26
		Medium	Private	183	55
			Public	54	15
		Low	Private	108	54
			Public	22	13
Female	Old	High	Private	82	17
			Public	85	16
		Medium	Private	46	16
			Public	60	11
		Low	Private	29	29
			Public	40	18
	Young	High	Private	160	23
			Public	152	28
		Medium	Private	89	17
			Public	56	21
		Low	Private	57	41
			Public	34	28

Use the procedure suggested in Section 4.5, Example 4.4 to select a graphical model for these data.

(a) Remove the lines in the association diagram in the following order:

(1) D-E (2) B-E (3) A-B (4) A-E (5) B-D (6) A-C

and write down for each step the possible graphical models obtained by removing each of the remaining lines in the next step.

(b) Make a list of z-test statistics and levels of significance explaining your selection procedure.

[Hint: Tables similar to those in Section 4.5 are shown in the Appendix.]

Chapter 5

Incomplete Tables

5.1 Random and structural zeros

A contingency table is incomplete if one or more cells have a zero count. We distinguish between **random** and **structural zeros**. If the cell count has expected value 0, i.e. the probability of observing an observation in the cell is 0, the zero a structural zero. If on the other hand the expected value, and thus the probability of an observation in a cell is larger than 0 an observed zero is random. Technically random and structural zeros are treated in the same way. The reason is that in any test statistic, a term corresponding to a cell with a zero count will cancel out, since all test statistics have the form

$$Z = 2 \sum \text{observed} \, (\ln(\text{observed}) - \ln(\text{expected})) \, .$$

If there are few zeros in a table the consequences are limited. Then only a few terms are missing in the Z-test statistic and we only need to compensate in the degrees of freedom for the test statistic used to test the fit of the model. The important thing is that the expected numbers under the model can be estimated. Since the likelihood equations all have the form of equating sufficient statistics with their expected values, problems only arise if one or more of the sufficient marginals for a model are zero. In this case there clearly will be no solutions to the likelihood equations. Only for the saturated model is it necessary that the table is complete with no zeros.

As an Example consider a 3-way contingency table under the model with sufficient marginals AB, AC. For this model the likelihood equations are

$$x_{ij.} = n\pi_{ij.} \, , \, i = 1,...,I \, , \, j = 1,...,J \tag{5.1}$$

and

$$x_{i.k} = n\pi_{i.k} \, , \, i = 1,...,I \, , \, k = 1,...,K \, . \tag{5.2}$$

If for Example $x_{111}, ... , x_{11k}$ are all zero, $x_{11.} = 0$ and according to (5.1) then also $n\pi_{ij.} = 0$. As expected this implies that there is no finite estimate for τ_{11}^{AB}. To see this, consider the likelihood equation (5.1) for i=1 and j=1, which since

$$\ln(n\pi_{11k}) = \tau_0 + \tau_1^A + \tau_1^B + \tau_k^C + \tau_{11}^{AB} + \tau_{1k}^{AC}, \quad (5.3)$$

and

$$n\pi_{11\cdot} = \sum_{k=1}^{3} n\pi_{11k}$$

becomes

$$x_{11\cdot} = n\pi_{11\cdot} = \exp\left(\tau_0 + \tau_1^A + \tau_1^B + \tau_{11}^{AB}\right) \sum_{k=1}^{3} \exp\left(\tau_k^C + \tau_{1k}^{AC}\right).$$

or

$$\ln(x_{11\cdot}) = \tau_0 + \tau_1^A + \tau_1^B + \tau_{11}^{AB} + \ln \sum_{k=1}^{3} \exp\left(\tau_k^C + \tau_{1k}^{AC}\right).$$

All parameters in this equation, except τ_{11}^{AB}, can be estimated with finite values, since all τ's, except τ_{11}^{AB}, correspond to non-zero sufficient marginals. Hence the equation can only be satisfied if the ML-estimate for τ_{11}^{AB} is $-\infty$. In this situation we say that τ_{11}^{AB} is non-estimable.

As a very simple Example of how the zero marginals determine which parameters are estimable, consider the hypothetical contingency table in Table 5.1.

TABLE 5.1. A hypothetical 2×2×3 table.

		k = 1	2	3
i = 1	j = 1	0	0	0
	2	0	43	7
2	j = 1	11	14	46
	2	13	21	5

In Table 5.1 there are 4 zeros. The zeros for i=1, j=1 and k = 1, 2, 3 may in this case be structural, while $x_{121} = 0$ may be random, but this does not influence the arguments in the following. There is only one marginal, which is zero, namely

$$x_{11\cdot} = 0. \quad (5.4)$$

This means that there is no estimated expected marginal $n\pi_{11\cdot}$, and no finite ML-estimate for the log-linear parameter τ_{11}^{AB}.

INCOMPLETE TABLES 129

If there is no estimate for τ_{11}^{AB}, the remaining τ_{ij}^{AB}'s are also non-estimable in this case with I = J = 2 in spite of the fact that $x_{12.} > 0$, $x_{21.} > 0$ and $x_{22.} > 0$. This follows from the constraints

$$\tau_{1.}^{AB} = \tau_{12}^{AB} + \tau_{11}^{AB} = 0,$$

$$\tau_{2.}^{AB} = \tau_{21}^{AB} + \tau_{22}^{AB} = 0$$

and

$$\tau_{.2}^{AB} = \tau_{12}^{AB} + \tau_{22}^{AB} = 0.$$

Thus also for log-linear parameters corresponding to non-null marginals, there may be non-estimable log-linear parameters due to the constraints defining the log-linear parameters. Example 5.1 in the next section gives more details on which parameters can be estimated in an incomplete table.

In summary: If for a given model all sufficient marginals are positive, all the log-linear parameters can be estimated and any zeros in the table only have the consequence of reducing the number of degrees of freedom for the Z-test statistic. This reduction is obtained by counting the number N of non-zero cells, subtracting 1 and subtracting the number of estimated unconstrained log-linear parameters. If for Example in Table 5.1 all the 3 zeros in the first row were changed to positive values, there would be N=11 non-zero cells and under the model AB, AC seven unconstrained log-linear parameters to be estimated (one τ^{AB}, two τ^{AC}'s, one τ^A, one τ^B and two τ^C's). Hence the Z-test statistic will have 11 terms and be approximately χ^2-distributed with df = 11 - 1 - 7 = 3 degrees of freedom. Note that this is one degree of freedom less than we would have found in a complete table, since under the model AB, AC there are two 3-factor interactions which are 0 and two τ^{BC}'s which are 0, giving 4 degrees of freedom.

5.2 Counting the number of degrees of freedom

As mentioned the number of degrees of freedom depends on the number of zeros in the contingency table in two ways. First, for any observed zero count in a cell the corresponding term in the Z-test statistic is missing and we must compensate in the number of degrees of freedom. Second, zero's may correspond to one or more sufficient marginals under the model being zero, in which case we must adjust for log-linear parameters not being estimable. Fortunately there is a formula for how to count the number of degrees of freedom correctly. This formula is based on the quantities N_0, $N_1(H)$ and $N_2(H)$ defined as

N_0 = Number of cells with observed count ≠ 0.
$N_1(H)$ = Number of unconstrained log-linear parameters under H in a complete table without zeros.

$N_2(H)$ = Number of unconstrained log-linear parameters under H for which the corresponding sufficient marginal is 0.

If Z(H) is the test statistic for H,

$$Z(H) \sim \chi^2(df(H)) ,$$

where

$$df(H) = N_0 - N_1(H) + N_2(H) . \qquad (5.5)$$

If we apply this formula to the data in Table 5.1 under the model AB, BC we get $N_0 = 8$ and $N_1(H) = 8$ since in a complete table the following log-linear parameters have to be estimated: τ_{11}^{AB}, τ_{11}^{BC}, τ_{12}^{BC}, τ_1^A, τ_1^C, τ_2^C, τ_1^B and τ_0. Finally $N_2(H) = 2$, since $x_{1.1} = 0$ and $x_{11.} = 0$. It follows that the number of degrees of freedom according to Equation (5.5) is

$$df(H) = 8 - 8 + 2 = 2.$$

In order to check that Equation (5.5) counts the number of degrees of freedom correctly, we use the result that the number of degrees of freedom is the number of unconstrained log-linear parameters set equal to zero. In this case no 3-factor interactions can be estimated. In the complete table there are two unconstrained 3-factor interactions τ_{111}^{ABC} and τ_{112}^{ABC}, but since both x_{111} and x_{112} are zero neither of these are estimable. Also the τ^{AB}'s are non-estimable, as we saw in section 5.1. In the model AB, BC only the 2-factor interactions τ_{ik}^{AC} are thus set to zero, and there are two unconstrained τ_{ik}^{AC}'s, such that

$$df(H) = 2 .$$

EXAMPLE 5.1. *Strenuous work.*
The data in Table 5.2 are extracted from the data base of the Danish Welfare Study. The sample is cross-classified according to the following 3 variables:

A: Strenuous work, with categories Yes, Yes sometimes and No.
B: Type of employment, with categories Blue collar employee, White collar employee and Employer.
C: Social group with 4 categories.

INCOMPLETE TABLES

TABLE 5.2. The Danish Welfare Study cross-classified according to Strenuous work, Type of employment and Social group.

A: Strenuous work	B: Type of employment	C: Social group			
		I-II	III	IV	V
Yes	Blue collar	0	0	64	182
	White collar	79	98	110	0
	Employer	38	126	19	0
Yes, sometimes	Blue collar	0	0	131	265
	White collar	156	166	292	0
	Employer	28	150	52	0
No	Blue collar	0	0	156	556
	White collar	136	166	382	0
	Employer	18	180	54	0

Source: The data base from the Danish Welfare Study. Cf. Example 3.2.

Table 5.3 shows the observed values $z(H)$ of the test statistic $Z(H)$ for four interesting models of types H_1 and H_2.

TABLE 5.3. The test statistic z(H) for four models applied to the data in Table 5.2

Model	z(H)	df
AB, AC, BC	43.66	4 (12)
AB, BC	92.43	10 (18)
AC, BC	116.07	8 (16)
AB, AC	194.16	6 (18)

The number of degrees of freedom in parentheses are the degrees of freedom for the model in a complete table. The reduction in the number of degrees of freedom is due to the zeros in the table. In this case the zeros are structural. They are a consequence of the way the social groups are constructed. Blue collar workers are always by definition in group IV or in group V. Hence there are structural zeros for $(j,k)= (1,1)$ and $(1,2)$. In addition neither white collar employees nor employers can be in social group V, since social group V is exclusively blue collar workers with no additional job education. This means that also cells with $(j,k) = (2,4)$ and $(3,4)$ are structural zeros. Thus the marginals $x_{.11}$, $x_{.12}$, $x_{.24}$ and $x_{.34}$ are all structural zeros. This influences which parameters are estimable. Variable A and B both have 3 levels, while variable C has 4 levels, which means that there are $2 \cdot 2 \cdot 3 = 12$ three-factor interactions which are 0 in all 4 models. Due to the structural zeros only some of these are estimable, however. The same is true for the $3 \cdot 2 = 6$ two-factor interactions between variables B and C. Table 4.4 provides a summary of the estimable parameters. In order to make the table easier to read the superscripts ABC and BC are omitted inside the table.

TABLE 5.4. The estimable three-factor interactions between A,B and C and the estimable two-factor interactions between B and C for the incomplete contingency Table 5.2.

τ_{ijk}^{ABC}		k = 1	2	3	4
i=1	j=1	-	-	*	*
	2	τ_{121}	τ_{122}	$-\tau_{121} - \tau_{122}$	-
	3	$-\tau_{121}$	$-\tau_{122}$	$\tau_{121} + \tau_{122}$	-
2	j=1	-	-	*	*
	2	τ_{221}	τ_{222}	$-\tau_{221} - \tau_{222}$	-
	3	$-\tau_{221}$	$-\tau_{222}$	$\tau_{221} + \tau_{222}$	-
3	j=1	-	*	*	*
	2	$-\tau_{121} - \tau_{221}$	$-\tau_{122} - \tau_{222}$	$\tau_{121} + \tau_{122}$ $+ \tau_{221} + \tau_{222}$	-
	3	$\tau_{121} + \tau_{221}$	$+\tau_{122} + \tau_{222}$	$-\tau_{121} - \tau_{122}$ $- \tau_{221} - \tau_{222}$	-

τ_{jk}^{BC}	k = 1	2	3	4
j=1	-	-	*	*
2	τ_{21}	τ_{22}	$-\tau_{21} - \tau_{22}$	-
3	$-\tau_{21}$	$-\tau_{22}$	$\tau_{21} + \tau_{22}$	-

In Table 5.4 parameters which are non-estimable due to structural zeros are marked by a "-". Parameters, which are non-estimable due to a normalization without the corresponding marginal being a structural zero are marked by a "". As an Example of a non-estimable parameter of the latter type, consider τ_{114}^{ABC}, which is non-estimable because both τ_{124}^{ABC} and τ_{134}^{ABC} correspond to marginals, which are structural zeros and*

$$\tau_{1.4}^{ABC} = 0.$$

From Table 5.4 we can conclude that the number of estimable 3-factor interactions is only 4, despite the fact that there are 12 unconstrained 3-factor interactions in a complete table. In the same way the table shows that there are only 2 estimable τ_{jk}^{BC}'s compared to the 6 τ_{jk}^{BC}'s in a complete table. It is now easy to count the correct number of degrees of freedom in Table 5.3. For the model AB, AC, BC only the 3-factor interactions between A, B and C are set to zero, so df = 4. For the model AB, BC in addition the 2-factor interactions between A and C are zero, but here are no structural zeros, so we add 6 degrees of freedom. In the same way we get df = 4 + 4 = 8 degrees of freedom for the model AC, BC. In the final model with sufficient marginals AB, AC in addition to the 3-factor interactions also the 2-factor interactions between B and C are zero, but since only two τ_{jk}^{BC}'s are estimable df = 4 + 2 = 6.

It is usually easier to apply Formula (5.5) than setting up and counting the τ's in a table like Table 5.4. For the model AC, BC we thus get $N_0 = 24$, $N_1(H) = 20$, because τ_0, 7 main effects, 6 τ_{ik}^{AC}'s and 6 τ_{jk}^{BC}'s add up to 20, and $N_2(H) = 4$, since there are 4 zeros in a table over x_{jk}, and none of these are a consequence of other zeros. The correct number of degrees of freedom is therefore

$$df = 24 - 20 + 4 = 8.$$

Note that for $N_2(H)$ we only count marginal zeros, corresponding to unconstrained parameters. For Example $x_{.14} = 0$ would not cause $N_2(H)$ to be increased by 1, since $x_{.14} = 0$ can be deduces from the 4 structural zeros.

5.3 Validity of the χ^2-approximation

In some contingency tables the problem is not structural or random zeros, but cells with small expected counts. According to section 2.6, the χ^2-approximations to the true percentiles of the distribution of the Z-test statistics are not necessarily valid if the cell probabilities are small. Since the expected counts are obtained as the cell probabilities multiplied by the sample size we should be careful when we evaluate the significance level of an observed Z-test statistic by the percentiles of a χ^2-distribution for small counts. There have been many investigations of the validity of the χ^2-approximations for the Z-test statistic. None of these give any conclusive answer to the question of a lower bound for the observed expected counts in order for the χ^2-approximation to be valid. Most studies agree, however, that the χ^2-approximation is valid if the expected counts are larger than 1, but if too many cells have expected counts close to 1, the χ^2-approximation is questionable. A sensible rule is to require that all expected counts are larger than 3.

If one or more cells have expected counts smaller than 3, one way to remedy the situation is to group cells in the computation of the Z-test statistic. A **grouping** of cells consists of adding both the observed and the expected counts for the cells in question. After the grouping the grouped observed and expected counts then only contribute one term to the Z-test statistic. This term will have the usual form

$$2 \cdot \text{observed} \cdot (\ln(\text{observed}) - \ln(\text{expected})),$$

but instead of several terms for cells with small expected counts there will be just one term. That the Z-test statistic after grouping is still approximately χ^2-distributed can be seen as follows: Let the contingency table be a 3-way table, and suppose that

$$n\hat{\pi}_{111} < 3, \; n\hat{\pi}_{112} < 3, \; n\hat{\pi}_{113} < 3.$$

The grouping could then be

$$\hat{\pi}^*_{111} = \hat{\pi}_{111} + \hat{\pi}_{112} + \hat{\pi}_{113}$$

and

$$x^*_{111} = x_{111} + x_{112} + x_{113}.$$

But if

$$(X_{111},...,X_{IJK}) \sim M(n,\pi_{111},...,\pi_{IJK}),$$

then

$$(X^*_{111}, X_{114},..., X_{IJK}) \sim M(n, \pi^*_{111}, \pi_{114},..., \pi_{IJK}),$$

where the dimension of the multinomial distribution is now IJK-2. Hence

$$Z = 2X^*_{111}\left(\ln X^*_{111} - \ln(n\hat{\pi}^*_{111})\right) + 2\sum_{ijk \neq (111, 112, 113)} X_{ijk}\left(\ln X_{ijk} - \ln(n\hat{\pi}_{ijk})\right)$$

is χ^2-distributed with IJK - 1 - m - 2 degrees of freedom, where m is the number of parameters in the model.

Whether we group or not, the degrees of freedom are always calculated as

df = number of terms in z(H) - 1 - number of estimated parameters.

Unfortunately the parameters have to reestimated since the model has changed. The model is thus not necessarily log-linear after grouping. We can hope that the ML-estimates for the parameters do not change too much when re-estimated, but this is not always the case.

Another way to account for small expected counts is to compute what are popularly known as **exact levels of significance**. This is a **Monte Carlo technique**, where we let a computer simulate a large number of contingency tables, for Example 500, which has the given total n and cell probabilities derived as the observed expected values divided by n. For each of these tables a Z-test statistic is computed and the percentage of the observed 500 z-values larger than the one observed in the original table is then an estimate of the true level of significance according to the law of large numbers.

EXAMPLE 5.2. *Leave schemes.*
In 1996 The Danish National Institute of Social Research carried out a large scale investigation of the effects of some new leave of absence schemes offered with government support to the Danish public. In Table 5.5 two of these schemes are compared with the variables:

> A: *Contact with substitute having the 9 categories, shown in the table, describing the way the employer got in contact with the substitute for the person on leave.*

B: Sector, with categories employed in the Private sector and employed in the Public sector

Variable C is the Leave scheme with categories Parental leave or Leave with the purpose of further education, here called Educational leave.

The symbol "AF" in the table stands for "Arbejdsformidlingen", the official Danish government agency with job centres in most cities, who advertises vacant jobs and offer job search facilities.

TABLE 5.5. A sample of 711 employers cross-classified according to method of Contact with substitute, Sector and the Leave scheme for the employee.

A: Contact with substitute	B: Sector	C: Leave scheme	
		Parental	Education
By the AF	Private	36	24
	Public	17	27
By advertisement	Private	37	20
	Public	74	43
Unannounced contact	Private	31	18
	Public	15	8
Employer has other substitute	Private	16	12
	Public	75	48
Recommended by other employee	Private	15	9
	Public	25	14
Recommended by the Union	Private	7	5
	Public	4	9
Recommended by City authorities	Private	0	1
	Public	9	10
Contact to substitute agency	Private	5	1
	Public	1	3
Other contact	Private	41	10
	Public	21	20

Source: Andersen, D., Appeldorn, A. and Weise, H.: Leave - an evaluation of the leave schemes. (In Danish). Report 96:11. The Danish National Institute of Social Research.

For this table a model with sufficient marginals AB and C barely fits this data with observed Z-value

$$z(H) = 32.37, \ df = 17$$

and level of significance

$$P(Q \geq 32.37) = 0.014.$$

The expected numbers under the model are shown in Table 5.6.

TABLE 5.6. Expected numbers for the data in Table 5.5 under the model AB, C.

		C: Leave scheme	
A: Contact	B: Sector	Parental	Education
By the AF	Private	36.2	23.8
	Public	26.5	17.5
By advertisement	Private	34.4	22.6
	Public	70.6	46.4
Unannounced contact	Private	29.6	19.4
	Public	13.9	9.1
Employer has other substitute	Private	16.9	11.1
	Public	74.2	48.8
Recommended by other employee	Private	14.5	9.5
	Public	23.5	15.5
Recommended by the Union	Private	7.2	4.8
	Public	7.8	5.2
Recommended by City authorities	Private	0.6	0.4
	Public	11.5	7.5
Contact to substitute agency	Private	3.6	2.4
	Public	2.4	1.6
Other contact	Private	30.8	20.2
	Public	24.7	16.3

We note that several of these values are small, in particular the values 0.6 and 0.4. In order to study the grouping procedure, the three expected counts smaller than

2 are merged into two combined "cells". First, cells (711), (712) and (722) are merged to give

$$x^*_{711} = 11 \, , \, n\hat{\pi}^*_{711} = 8.5 \, .$$

Second, cells (812) and (822) are merged to give

$$x^*_{812} = 4 \, , \, n\hat{\pi}^*_{812} = 4.0 \, .$$

The procedure is now to omit the five terms in z(H) corresponding to the merged cells, but then to add the two terms

$$2 \cdot \{ \, 11 \cdot (\ln(11) - \ln(8.5)) \, \} + 2 \cdot \{ \, 4 \cdot (\ln(4) - \ln(4.0)) \, \}$$

to z(H). Doing this we get

$$z(H) = 28.42 \, , \, df = 14$$

with level of significance

$$P(Q \geq 28.42) = 0.013.$$

Note that the grouping causes a reduction in the number of degrees of freedom. The result of the grouping seems to indicate, that in this case, the χ^2-approximation is valid. If we simulate an exact level of significance based on 500 simulated tables with the expected values in Table 5.6 the result is

$$P(Z \geq 32.37) = 0.017,$$

confirming that in this case the χ^2-approximation indeed seems to work.

If possible we should try to compute simulated exact levels of significance. But since it is time consuming, we may have to rely on grouping procedures.

5.4 Exercises

5.1 Consider the 3-way table

		k = 1	2	3
i = 1	j = 1	0	0	8
	2	5	43	7
2	j = 1	0	0	46
	2	8	21	5

(a) Count the number of degrees of freedom for the model AB, AC, BC.

(b) Count the number of degrees of freedom for the models AB, AC, and AB, BC.

5.2 Consider the 3-way table

		k = 1	2	3
i = 1	j = 1	42	33	11
	2	0	0	4
	3	11	12	8
2	j = 1	53	58	46
	2	0	0	5
	3	15	8	11
3	j = 1	26	18	5
	2	0	0	4
	3	14	17	6

(a) Count the number of degrees of freedom for the model AB, AC, BC by Formula (5.5).

(b) Count the number of degrees of freedom for the models (1) AB, AC, (2) AB, BC and (3) AC,BC by Formula (5.5).

(c) Check the results in (a) and (b) by making tables like those in Table 5.4.

5.3 Suppose we want to compare time spent on training per for four different sports disciplines: Track and field, tennis, boxing and wrestling for both men and women athletes. In boxing and wrestling there are very few practitioners, and a few years ago, none. Hence a 3-way table may look like this (the data are fictitious).

Discipline	Sex	Hours spent on training per day		
		1	1-2	3 or more
Track & Field	Male	25	102	32
	Female	33	89	25
Tennis	Male	12	99	44
	Female	14	75	43
Boxing	Male	24	78	17
	Female	0	0	0
Wrestling	Male	22	23	15
	Female	0	0	0

(a) Count the number of degrees of freedom for models AB, AC and AC and BC, when variable A is Discipline, variable B is Sex and variable C is Training hours.

(b) Compare the numbers in (a) with the degrees of freedom one gets in a complete table.

(c) Show in a table that there is, in fact, only one 3-factor interaction in this incomplete table. [Hint: Take inspiration from Table 5.4.]

Chapter 6

The Logit Model

6.1 The logit model

In many contingency tables the aim of the analysis is to explain the variation in one of the variables by the variation of other variables. Such a special variable of interest is called a **response variable**. Those variables in the contingency table, which explain the variation in the response variable, are called **explanatory variables**.

The assumptions for the logit model, to be introduced below, are

(i) The response variable is binary.
(ii) The statistical model for the contingency table formed by the response variable and the explanatory variables is log-linear.

As an introduction to the logit model consider the model

$$\left(X_{1\ldots 1},\ldots,X_{I_1\ldots I_m}\right) \sim M\left(n;\pi_{1\ldots 1},\ldots,\pi_{I_1\ldots I_m}\right), \quad (6.1)$$

for an m-dimensional contingency table such that

$$E\left[X_{i_1\ldots i_m}\right] = n\pi_{i_1\ldots i_m}.$$

For $i_1 = 1$ we get according to assumption (ii)

$$\ln \pi_{1 i_2 \ldots i_m} = \tau_0^* + \tau_1^A + \tau_{i_2}^B + \ldots + \tau_{i_m}^S + \tau_{1 i_2}^{AB} + \ldots + \tau_{i_{m-1} i_m}^{RS} + \ldots + \tau_{1 i_2 \ldots i_m}^{AB\ldots S}. \quad (6.2)$$

For $i_2 = 2$ we get in the same way

$$\ln \pi_{2 i_2 \ldots i_m} = \tau_0^* + \tau_2^A + \tau_{i_2}^B + \ldots + \tau_{i_m}^S + \tau_{2 i_2}^{AB} + \ldots + \tau_{i_{m-1} i_m}^{RS} + \ldots + \tau_{2 i_2 \ldots i_m}^{AB\ldots S}. \quad (6.3)$$

From (6.2) and (6.3) then follows that

$$\ln(\pi_{1i_2\ldots i_m}) - \ln(\pi_{2i_2\ldots i_m})$$
$$= \tau_1^A - \tau_2^A + \tau_{1i_2}^{AB} - \tau_{2i_2}^{AB} + \ldots + \tau_{1i_2\ldots i_m}^{AB\ldots S} - \tau_{2i_2\ldots i_m}^{AB\ldots S} \,,$$
(6.4)

since all interactions, which do not have A as a subscript, cancel out. Due to the constraints

$$\tau_1^A + \tau_2^A = 0, \quad \tau_{1i_2}^{AB} + \tau_{2i_2}^{AB} = 0 \,, \text{ etc.},$$

it follows from (6.4) that

$$\ln(\pi_{1i_2\ldots i_m}) - \ln(\pi_{2i_2\ldots i_m}) = 2\tau_1^A + 2\tau_{1i_2}^{AB} + \ldots + 2\tau_{1i_2\ldots i_m}^{AB\ldots S} \,, \quad (6.5)$$

The **prediction probability** of observing the response variable A at level 1 given the explanatory variables is defined as

$$\pi_{1 \mid i_2 \ldots i_m} = P\big(A = 1 \mid \text{levels } i_2 \ldots i_m\big) \,. \quad (6.6)$$

Here the symbol "A=1 | levels $i_2 \ldots i_m$" stands for: Variable A is observed at level 1 given that the explanatory variables B, C, ... S are observed at levels $i_2, \ldots, i_m$. The connection between the prediction probability (6.6) and the expression (6.5) becomes clearer, when we introduce the **logit function**

$$y = \text{logit}(x) = \ln\left(\frac{x}{1-x}\right), \quad (6.7)$$

for which the inverse function is

$$x = \frac{\exp(y)}{1 + \exp(y)} \,.$$

The logit-function (6.7) is shown in Figure 6.1

THE LOGIT MODEL

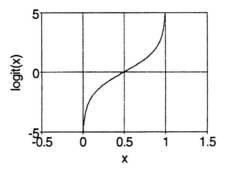

FIGURE 6.1. The logit function.

Since

$$\pi_{1|i_2...i_m} = \frac{\pi_{1i_2...i_m}}{\pi_{1i_2...i_m} + \pi_{2i_2...i_m}}$$

and

$$1 - \pi_{1|i_2...i_m} = \frac{\pi_{2i_2...i_m}}{\pi_{1i_2...i_m} + \pi_{2i_2...i_m}},$$

it follows from (6.5) that

$$\text{logit}(\pi_{1|i_2...i_m}) = 2\tau_1^A + 2\tau_{1i_2}^{AB} + ... + 2\tau_{1i_2...i_m}^{AB...S}, \quad (6.8)$$

Figure 6.1 shows that logit(x) has the limit $+\infty$ when x tends to 1 and the limit $-\infty$ when x tends to 0. The prediction probability thus has range (0,1), while the logit-transformed prediction probability has range $(-\infty,+\infty)$. It follows from (6.8) that if the value of the logit (6.8) is large then the probability of observing variable A at level 1 is much larger than the probability of observing variable A at level 2. In the same way if the logit is large negative, then we are more likely to observe variable A at level 2.

We further note that the logit (6.8) only depends on those interactions involving the response variable A.

Finally (6.8) shows that the magnitudes and signs of the interactions between the response variable and the explanatory variables determine if we are more likely to get a response 1 or a response 2 on variable A. If interactions with positive values dominate, the response is more likely to be 1 and if interactions with negative values dominate, the response is more likely to be 2. If, however, the sum of the interactions are 0 then the probability of response 1 and the probability of response 2 are equal and, therefore, 1/2. Note that there is no constant term τ_0 in (6.8), so that it is the balance between positive and negative interactions with variable A that determines the prediction probability.

Two features of the logit model are important:

(i) The estimation problem is already solved by the methods discussed in chapters 3 and 4. In fact the ML-estimates for the interactions are those obtained from an analysis of a log-linear model for the contingency table formed by all the variables, under the restrictions imposed by the logit model.

(ii) Only interactions which involve the response variable A are part of the model. Hence we do not need to estimate interactions between the explanatory variables.

On the other hand, if we want to obtain the ML-estimates from a log-linear model for the full contingency table, the correct model must include all interactions up to the highest dimension between the explanatory variables.

6.2 Hypothesis testing in the logit model

In this section we consider only cases with three explanatory variables. (Extensions to more explanatory variables are straight forward.) In this case the logit model has the form

$$\text{logit}(\pi_{1|jkl}) = 2\tau_1^A + 2\tau_{1j}^{AB} + 2\tau_{1k}^{AC} + 2\tau_{1l}^{AD}$$
$$+ 2\tau_{1jk}^{ABC} + 2\tau_{1jl}^{ABD} + 2\tau_{1kl}^{ACD} + 2\tau_{1jkl}^{ABCD} .$$
(6.9)

This model, describing the influence of the explanatory variables on the response variables, can be considerably simplified if some of the interactions involving more than one explanatory variable can be omitted. Of special interest are those cases where the logit model (6.9) contains only 2-factor interactions. One interesting basic hypothesis is accordingly

$$H_0 : \tau_{1jk}^{ABC} = \tau_{1jl}^{ABD} = \tau_{1kl}^{ACD} = \tau_{1jkl}^{ABCD} = 0 \tag{6.10}$$

for all j, k and l.

Since interactions between explanatory variables are not specified in a logit model,

THE LOGIT MODEL

the sufficient marginals for H_0 are

$$AB, AC, AD, BCD.$$

A test for H_0 is, therefore, identical with a test for a log-linear model with sufficient marginals AB, AC, AD and BCD against the saturated model in the contingency table formed by all four variables A, B, C and D.

If H_0 can be accepted, the logit model reduces to

$$\text{logit}(\pi_{1|jkl}) = 2\tau_1^A + 2\tau_{1j}^{AB} + 2\tau_{1k}^{AC} + 2\tau_{11}^{AD}. \tag{6.11}$$

In (6.11) the influence of the explanatory variables is expressed directly in the 2-factor interactions τ_{1j}^{AB}, τ_{1k}^{AC} and τ_{11}^{AD}. Thus if $\tau_{11}^{AD} = 0$ for all l, variable D does not contribute to the description of the variation in the response variable A. Hence given that the logit model (6.11) describes the data, we can evaluate the contribution of each of the explanatory variables either by comparing the three parallel hypotheses

$$H_1: \tau_{11}^{AD} = 0, \, l = 1,\ldots,L,$$

$$H_2: \tau_{1k}^{AC} = 0, \, k = 1,\ldots,K$$

and

$$H_3: \tau_{1j}^{AB} = 0, \, j = 1,\ldots,J,$$

using their level of significance in a test against H_0, or by evaluating the level of significance for the three sequential hypotheses

$$H_{(1)} = H_1: \tau_{11}^{AD} = 0, \, l = 1,\ldots,L,$$

$$H_{(2)}: \tau_{1k}^{AC} = \tau_{11}^{AD} = 0, \, k = 1,\ldots,K, \, l = 1,\ldots,L$$

and

$$H_{(3)}: \tau_{1j}^{AB} = \tau_{1k}^{AC} = \tau_{11}^{AD} = 0, \, j = 1,\ldots,J, \, k = 1,\ldots,K, \, l = 1,\ldots,L,$$

where $H_{(1)} = H_1$ is compared with H_0, $H_{(2)}$ with $H_{(1)}$ and $H_{(3)}$ with $H_{(2)}$. Of course a sequential procedure is only valid if variable D is the most likely to be omitted from the model, variable C the second most likely and variable B the least likely to be omitted.

As mentioned, each logit model is equivalent to a log-linear model for the contingency table between variables A, B, C and D. Each of the hypotheses H_0, $H_{(3)}$, $H_{(2)}$, $H_{(1)}$, H_3, H_2 and H_1 thus corresponds to a model given by its sufficient marginals, where the 3-factor interaction and all lower dimension interactions between B, C and D are always included. Table 6.1 gives a summary of the sufficient marginals for the different hypotheses and the corresponding logit models. The table also shows the number of degrees of freedom, when the model is tested against the full logit model (6.9).

TABLE 6.1. The hypotheses H_1, H_2, H_3, $H_{(2)}$ and $H_{(3)}$ for a logit model and the corresponding sufficient marginals and the number of degrees of freedom when tested against the saturated model.

Hypothesis	Omitted variable	Sufficient marginals	Number of degrees of freedom
H_0	-	AB, AC, AD, BCD	$(I - 1)(JKL - J - K - L + 2)$
H_3	B	AC, AD, BCD	$(I - 1)(JKL - L - K + 1)$
H_2	C	AB, AD, BCD	$(I - 1)(JKL - L - J + 1)$
$H_1 = H_{(1)}$	D	AB, AC, BCD	$(I - 1)(JKL - J - K + 1)$
$H_{(2)}$	C, D	AB, BCD	$(I - 1)(JKL - J)$
$H_{(3)}$	B, C, D	A, BCD	$(I - 1)(JKL - 1)$

From Table 6.1 it is easy to form a table of tests from which the most simplest logit model, which fits the data to a satisfactory degree, can be selected. The selected logit model then tells us which explanatory variable must be included in the model. If for example hypothesis H_1 holds, the analysis has shown that D can be omitted as explanatory variable, but not B and C. The logit model is then

$$\text{logit}(\pi_{1|jkl}) = 2\tau_i^A + 2\tau_{1j}^{AB} + 2\tau_{1k}^{AC}. \tag{6.12}$$

After it has been determined which variables are necessary to describe the variation in the response variable, there are two ways to report the results of an analysis by a logit model:

(1)   The influence of the explanatory variables can be described through the signs and the magnitudes of the 2-factor interactions included in the model.

As we saw in connection with (6.8), the larger the value of the logit, the larger the likelihood of response 1 and the smaller the value of the logit - i.e the larger a negative value - the larger the likelihood of response 2 on the response variable. If thus the estimated value of τ_{1j}^{AB} for a certain category j is positive, persons with variable B observed in category j will be more likely to have a response 1 than if τ_{1j}^{AB} has the value 0. For a negative estimated value of τ_{1j}^{AB} a response 2 will, in the same way, be more likely than if τ_{1j}^{AB} has the value 0. The absolute value of the 2-factor interactions may depend on many features of the data setup. It is, therefore, desirable to have **standardized estimates**, which scale the estimates to have the same expected range as a standardized normal deviate. Such standardized estimates are computed as

THE LOGIT MODEL 147

$$\hat{\omega}_{1j}^{AB} = \frac{\hat{\tau}_{1j}^{AB}}{\sqrt{\hat{var}\left[\hat{\tau}_{1j}^{AB}\right]}}. \quad (6.13)$$

Since the estimates in (6.13) are standardized, they provide us with measures of the **relative strength** of the influence of explanatory variable B on the response variable.

(2) For a given combination of observed categories of the explanatory variables we can estimate the **prediction probability**, which according to (6.8) is

$$\hat{\pi}_{1|jkl} = \frac{\exp(\hat{g}_{jkl})}{1 + \exp(\hat{g}_{jkl})}, \quad (6.14)$$

where, under H_1 for example,

$$\hat{g}_{jkl} = \text{logit}(\hat{\pi}_{1|jkl}) = 2\hat{\tau}_1^A + 2\hat{\tau}_{1j}^{AB} + 2\hat{\tau}_{1k}^{AC}, \quad (6.15)$$

We can thus estimate the prediction probability $\hat{\pi}_{1|jkl}$ if we have estimated values of the τ's. A list of the prediction probabilities $\hat{\pi}_{1|jkl}$ for all possible combinations of j, k and l in the contingency table can provide us with an overview of the way the explanatory variables influence the expected response on the response variable. Note that what terms we include on the right hand side in (6.15) will depend on which of the hypotheses H_3, H_2, H_1, $H_{(2)}$ or $H_{(3)}$ has been accepted. Hence the prediction probability need not be calculated for different values of l, if variable D does not contribute to the description of the variation in the response variable.

According to Theorem 2.2 the estimates of the log-linear parameters are approximately normally distributed with a covariance matrix, which is easy to estimate. It follows that we can establish 95% confidence limits for the prediction probability. In fact the confidence limits for the logit (6.15) are

$$g_{jkl} = \text{logit}(\pi_{1|jkl}) \in \hat{g}_{jkl} \pm 1.96\hat{\sigma}_{jkl}, \quad (6.16)$$

where

$$\hat{\sigma}_{jkl}^2 = \hat{var}[\hat{g}_{jkl}] = 4\hat{var}[\hat{\tau}_1^A] + 4\hat{var}[\hat{\tau}_{1j}^{AB}] + 4\hat{var}[\hat{\tau}_{1k}^{AC}]$$
$$+ 8\hat{cov}[\hat{\tau}_1^A, \hat{\tau}_{1j}^{AB}] + 8\hat{cov}[\hat{\tau}_1^A, \hat{\tau}_{1k}^{AC}] + 8\hat{cov}[\hat{\tau}_{1j}^{AB}, \hat{\tau}_{1k}^{AC}], \quad (6.17)$$

and all the variances and covariances in (6.17) are elements of the covariance matrix for the ML-estimates of the τ's. From (6.16) confidence limits for the prediction probabilities are obtained by transforming the limits by the inverse function to the logit function (6.7). In this way we get

$$\pi_{1|jkl} \in \left(\frac{\exp(g^-)}{1+\exp(g^-)}, \frac{\exp(g^+)}{1+\exp(g^+)} \right) \quad (6.18)$$

where

$$g^+ = \hat{g}_{jkl} + 1.96\hat{\sigma}_{jkl} \text{ and } g^- = \hat{g}_{jkl} - 1.96\hat{\sigma}_{jkl}.$$

Most computer programs will provide the estimated prediction probabilities $\hat{\pi}_{1|jkl}$, as well as the confidence limits (6.18) as output options.

EXAMPLE 6.1. *Party membership.*
In this example we shall study factors influencing membership of a political party in Denmark. The data are drawn from the data base of the Danish Welfare Study 1976. The response variable is thus

> A: Member of a political party, with response categories Yes or No.

The explanatory variables are

> B: Sex, with categories Women and Men.
> C: Employment sector, with categories Public sector and Private sector.
> D: Residence, with categories Living in Copenhagen or Living outside Copenhagen.

The resulting contingency table is shown as Table 6.2.

TABLE 6.2. The Danish Welfare Study cross-classified according to Membership of a political party, Sex, Employment sector and Residence.

A: Membership of party	B: Sex	C: Employment sector	D: Residence in Copenhagen	
			Yes	No
Yes	Women	Public	12	31
		Private	5	20
	Men	Public	16	37
		Private	19	73
No	Women	Public	175	375
		Private	162	475
	Men	Public	111	266
		Private	241	906

Source: The data base of the Danish Welfare Study. Cf. Example 3.2.

In order to determine the simplest logit model which fits the data, we compute the test statistics for the hypotheses in Table 6.1. The result is shown in Table 6.3,

The Logit Model

which shows both the tests $z(H)$ for H against H_0 and the sequential tests $z(H/H^*)$. For the sequential tests, H^* is H_0 for the tests of H_1, H_2, and H_3, while H^* is $H_{(1)}$ for the test of $H_{(2)}$, and $H_{(2)}$ for the test of $H_{(3)}$.

TABLE 6.3 Test statistics for different logit models fitted to the data in Table 6.2.

H	Minus var.	Sufficient marginals	z(H)	df	p	z(H|H*)	df	p
H_0		AB,AC, AD,BCD	0.64	4	0.958	-	-	-
H_3	B	AC,AD,BCD	18.48	5	0.002	17.84	1	0.000
H_2	C	AB,AD,BCD	17.12	5	0.004	16.48	1	0.000
$H_1=H_{(1)}$	D	AB,AC,BCD	0.87	5	0.973	0.23	1	0.028
$H_{(2)}$	C,D	AB,BCD	17.14	6	0.009	16.13	1	0.000
$H_{(3)}$	B,C,D	A,BCD	28.80	7	0.000	31.66	1	0.000

The first line in Table 6.3 shows that the logit model (6.11) fits the data to a satisfactory degree. The next three lines show that only H_1 can be accepted, i.e. Sex and Employment sector influence Party membership, while persons living in Copenhagen are as often party members as persons living outside Copenhagen, given the two other explanatory variables.

Method (1) of reporting the results of the analysis by a logit model is shown in Table 6.4.

TABLE 6.4. ML-estimates (Est.) and standardized estimates (Std. est.) for the significant interactions in the logit model (6.11).

$\hat{\tau}^{AC}$:		Employment sector	
Membership of party		Public	Private
Yes	Est.	0.302	-0.302
	Std.est.	8.158	-8.158
No	Est.	-0.302	0.302
	Std.est.	-8.158	8.158

$\hat{\tau}^{AB}$:		Sex	
Membership of party		Women	Men
Yes	Est.	-0.324	0.324
	Std.est.	-8.260	8.260
No	Est.	0.324	-0.324
	Std.est.	8.250	-8.260

From this table we can conclude that given other variables are constant more persons in the public sector than in the private sector are members of a political party, and that more men than women are members of a political party.

Method (2) of reporting the results of the analysis by a logit model is shown in Table 6.5.

TABLE 6.5. Prediction probabilities for Party membership given the different combinations of Sex and Employment sector.

B: Sex	C: Employment sector	Prediction probabilities for party membership
Women	Public	0.070
	Private	0.040
Men	Public	0.126
	Private	0.073

The prediction probabilities confirm the conclusions from Table 6.4.

6.3 Logit models with higher order interactions

The logit model (6.9) is equivalent to the saturated model for the contingency table for all variables, while the logit model (6.11) includes only 2-factor interactions. Between these two models there are a number of models of interest, although they can not be treated in as simple a way as the model with only 2-factor interactions. For example, models with a few 3-factor interactions included can in some cases be given useful interpretations. We consider, as an example, the following logit model with one set of three-factor interactions

$$\text{logit}(\pi_{1|jkl}) = 2\tau_1^A + 2\tau_{1j}^{AB} + 2\tau_{1k}^{AC} + 2\tau_{1l}^{AD} + 2\tau_{1jk}^{ABC} .$$

This model corresponds to the hypothesis

$$H_{01}: \tau_{1j1}^{ABD} = \tau_{1k1}^{ACD} = \tau_{1jkl}^{ABCD} = 0 , \text{ for all } j, k \text{ and } l .$$

Here the logit model depends on the terms τ_{1j}^{AB}, τ_{1k}^{AC} and τ_{1l}^{AD}, expressing the direct influence of the three explanatory variables, and on the three-factor interaction τ_{1jk}^{ABC}, which is an expression for the **joint influence** of the explanatory variables B and C. Such a joint influence of two variables can manifest itself by the prediction probability being high if the 3-factor interaction between A, B and C is large positive (all other interactions having neutral values) or by the prediction probability being low if the 3-factor interaction between A, B and C is large and negative.

EXAMPLE 6.2. *Alcohol consumption.*
Table 6.6 shows the contingency table formed by a cross-classification of the persons in the Danish Welfare Study 1976 with respect to the response variable

   A: *Alcohol consumption, with categories 0-1 units a day and 2 or more units a day.*

(*A unit is roughly one 33 cl bottle of beer of strength 4% vol. or its equivalent.*)

The explanatory variables are

   B: *Social group with four categories (cf. Example 1.1).*
   C: *Marriage status, with categories Married and Not married.*
   D: *Age with categories 20-39 years old and 40-69 years old.*

(*According to current Danish terminology "married" includes "living in a permanent partnership".*)

TABLE 6.6. The Danish Welfare Study 1976 cross-classification with respect to Alcohol consumption, Social group, Marriage status and Age.

A: Alcohol consumption	B: Social group	C: Marriage status	D: Age 20-39	D: Age 40-69
0-1 units a day	I-II	Married	43	38
		Not married	11	6
	III	Married	112	158
		Not married	35	33
	IV	Married	195	169
		Not married	76	44
	V	Married	135	216
		Not married	72	61
More than 2 units a day	I-II	Married	149	134
		Not married	37	28
	III	Married	200	225
		Not married	80	45
	IV	Married	296	238
		Not married	195	50
	V	Married	131	212
		Not married	130	49

Source: The data base of the Danish Welfare Study. Cf. Example 3.2.

In this case a logit model with only 2-factor interactions between the response variable and the explanatory variables does not fit the data. The Z-test statistic for the model with sufficient marginals AB, AC, AD, BCD, i.e the logit model

$$\text{logit}(\pi_{1|jkl}) = 2\tau_1^A + 2\tau_{1j}^{AB} + 2\tau_{1k}^{AC} + 2\tau_{11}^{AD}, \qquad (6.19)$$

has observed value

$$z(H) = 18.81, \ df = 10$$

with level of significance 0.043. The fit is, therefore, not quite good enough. As expected it is the 3-factor interaction between variables A, C and D which can not be omitted. When we add this interaction to the logit model (6.19), we get the logit model

$$\text{logit}(\pi_{1|jkl}) = 2\pi_1^A + 2\pi_{1j}^{AB} + 2\pi_{1k}^{AC} + 2\pi_{11}^{AD} + 2\pi_{1kl}^{ACD}. \qquad (6.20)$$

The ML-estimates of the parameters are obtained by estimating the log-linear parameters for a model with sufficient marginals

AB, ACD, BCD .

THE LOGIT MODEL 153

The observed value of the Z-test statistic for this model is

$$z(H) = 5.93 \, , \, df = 9$$

with level of significance

$$p = P(Q \geq 5.93) = 0.747.$$

The prediction probability for alcohol consumption thus depends on the explanatory variables in a slightly more complicated way. Since response 1 on the response variable means a low alcohol consumption, the estimated parameters in Table 6.7 show that alcohol consumption is lower in the lower social groups than in the two highest social groups. The table also shows that alcohol consumption is lower for married people than for unmarried, and lower for old people than for young people. But on top of this we must add the contributions from the estimated 3-factor interactions τ^{ACD}, where the estimates in Table 6.7 show that the alcohol consumption is lower for the combinations married-young and unmarried-old than for the combinations married-old and unmarried-young.

It is obvious from these somewhat complicated conclusions that it is in general preferable to try to avoid logit models with 3-factor interactions.

TABLE 6.7. Estimated parameters in the logit model (6.20).

Social group	I-II	III	IV	V
$2\hat{\tau}^{AB}$:	-0.716	0.070	0.132	0.514

Marriage status	Married	Unmarried
$2\hat{\tau}^{AC}$:	0.084	-0.084

Age	20-39	40-69
$2\hat{\tau}^{AD}$:	-0.190	0.190

	Marriage status	Age 20-39	40-69
$2\hat{\tau}^{ACD}$:	Married	0.148	-0.148
	Unmarried	-0.148	0.148

If we report the results by method (2), we get the prediction probabilities in Table 6.8.

TABLE 6.8. Prediction probabilities for the logit model (6.20).

B: Social group	C: Marriage status	D: Age	$\hat{\pi}_{1|jkl}$
I-II	Married	20-39	0.219
		40-69	0.233
	Not married	20-39	0.150
		40-69	0.257
III	Married	20-39	0.382
		40-69	0.402
	Not married	20-39	0.280
		40-69	0.433
IV	Married	20-39	0.395
		40-69	0.415
	Not married	20-39	0.291
		40-69	0.447
V	Married	20-39	0.489
		40-69	0.510
	Not married	20-39	0.376
		40-69	0.542

From Table 6.8 we can deduce which groups in the 1974 Danish population, when the data was collected, had a relative low alcohol consumption (high prediction probability in Table 6.8), for example unmarried old people in social group V. A group with a relatively high alcohol consumption are young, unmarried people in the highest social groups.

6.4 The logit model as a regression model

The logit model (6.8) can be formulated as a regression model, although such a model takes different forms depending on two things:

(i)   The number of categories for the explanatory variables.

(ii)   The dimension of the interactions between the response variable and the explanatory variables.

Here we only consider case (i) with binary explanatory variables, i.e. when the possible values of j, k and l are all 1 or 2. For this case we introduce the regression parameters β_0, β_1, β_2 and β_3 as

$$2\pi_1^A = \beta_0,$$

The Logit Model

$$2\pi_{11}^{AB} = \beta_1,$$

and

$$2\pi_{11}^{AC} = \beta_2,$$

$$2\pi_{11}^{AD} = \beta_3.$$

Then (6.11), which is the logit model with only 2-factor interactions, becomes

$$\text{logit}(\pi_{1|jkl}) = \beta_0 + \beta_1 z_j + \beta_2 z_k + \beta_3 z_l, \quad k,j,l = 1,2 \qquad (6.21)$$

where $z_1 = 1$ and $z_2 = -1$. Formula (6.21) is true given (6.11) for both j, k and l equal to 1 and 2, since $\tau_{11}^{AB} + \tau_{12}^{AB} = 0$ implies that $\tau_{11}^{AB} = -\tau_{12}^{AB}$ and we for $j = 1$ get

$$2\tau_{11}^{AB} = \beta_1 \cdot 1 = \beta_1,$$

and for $j = 2$ get

$$2\tau_{11}^{AB} = \beta_1(-1) = -\beta_1 = -2\tau_{12}^{AB}.$$

With a "scoring" +1 and -1 of the binary categories for the explanatory variables, the logit-transformed response probability (6.21) for response $i = 1$ given the explanatory variables thus becomes a linear regression model. As

$$\text{logit}(\pi_{2|jkl}) = -\text{logit}(\pi_{1|jkl}),$$

the same is the true for the logit-transformed response probability for response $i = 2$.

Formulated as a regression model, the hypotheses H_1, H_2 and H_3 in section 6.2 has a new interpretation.

$$H_1: \tau_{11}^{AD} = 0, \, l = 1,2$$

thus becomes

$$H_1: \beta_3 = 0,$$

and the sequential procedure in Table 6.1, where first variable D is omitted, then C and finally B, corresponds to a sequential procedure in a regression analysis, where the hypotheses $\beta_3 = 0$, $\beta_2 = 0$ and $\beta_1 = 0$ are tested sequentially.

6.5 Bibliographical notes

The idea of using the logistic function to transform probabilities goes far back. As a statistical tool for modelling data it seems to have been suggested first by Berkson (1973) in connection with so-called bioassay data. As a regression model it was introduced by Cox (1970). The name logit model has been used for different models. Since analysis by the regression model is now widely known as logistic regression, we have in this book followed a recent tradition to reserve the name logit model for models with binary responses and categorical explanatory variables where there is a direct connection to the log-linear models for contingency tables.

The logit model is covered in most basic textbooks on categorical data, for example Andersen (1990), chapter 8, or Agresti (1996) chapter 4.

6.6 Exercises

All the exercises for this chapter takes the form of a reanalysis of the exercises in chapters 4 and 5. The questions to answer are the same:

(a) Reformulate the model as a logit model with variable ... as the response variable and the remaining variables as explanatory variables.

(b) Estimate the regression parameters of the logit model.

(c) For data with only binary response variables estimate the parameters of the regression model.

6.1 Exercise 3.7 with variable A as response variable.

6.2 Exercise 3.8 with variable A as response variable.

6.3 Exercise 3.10 with variable C as response variable.

6.4 Exercise 4.1 with variable A as response variable. (Pick one of the two tables).

6.5 Exercise 4.3 with variable D as response variable.

6.6 Exercise 4.4 with variable C as response variable.

6.7 Exercise 4.5 with variable D as response variable.

6.8 Exercise 4.7 with variable A as response variable.

6.9 Exercise 4.8 with variable E as response variable.

Chapter 7

Logistic Regression Analysis

7.1 The logistic regression model

In section 6.3 it was demonstrated how the logit-model can be written as a regression model. In this chapter we treat regression models for categorical response variables in more details. A common name for these models is the **logistic regression model** because the regression part of the models, i.e. a linear combination of the values of the explanatory variables and the regression coefficients, is a **logistic transformation** of the probabilities of the response categories. The logistic transformation is given by the function

$$y = \ln\left(\frac{x}{1-x}\right),$$

discussed in section 6.1 and shown in Figure 6.1. The usefulness of this transformation lies in the fact, that it transforms the interval between 0 and 1 on to the real axis $(-\infty, +\infty)$. If we, therefore, transform probabilities by the logistic transformation, the probabilities will be "stretched" out over the complete real axis. A linear regression model directly for a response probability π, for example

$$\pi = \beta_0 + \beta_1 x ,$$

would cause the predicted value of the probability to be outside the permissible interval (0,1) if x becomes large positive or negative. This predictable performance can be avoided if we transform a response probability by the logistic transformation, so that we instead have

$$\text{logit}(\pi) = \ln\left(\frac{\pi}{1-\pi}\right) = \beta_0 + \beta_1 x .$$

Now the range is $(-\infty, +\infty)$ and large positive values of the term $\beta_1 x$ will predict probabilities near 1, and large negative values of $\beta_1 x$ will predict probabilities near 0, without going outside the range of π.

The basic logistic regression model deals with the binary case, where the range of the response variable consists of just two values. In section 7.9 we discuss situations with more than two response categories, so-called polytomous response variables. The logistic regression model in addition assumes that we have N independent joint observations of the **response variable** y and the **explanatory variables** $z_1,...,z_p$. Thus let

$$(y_v, z_{1v}, ..., z_{pv}), v = 1, ..., N$$

be the jointly observed values of the response variable y and the explanatory variables. The vector $(y_v, z_{1v}, ..., z_{pv})$ for a given value of v is called a **case**. In the model the z's are regarded as known values without random variation, while the random variable Y_v, corresponding to the observed value y_v, has possible values 0 and 1, 1 and 2 or "Yes" and "No". It the following we shall for consistency in the notation always use 0 and 1. The **response probability**

$$\pi_v = P(Y_v = 1 | z_{1v}, ..., z_{pv})$$

is a linear regression model in the z's after the response probability π_v has been transformed logistically, i.e.

$$\ln\left(\frac{\pi_v}{1-\pi_v}\right) = \beta_0 + \sum_{j=1}^{p} \beta_j z_{jv} . \tag{7.1}$$

This corresponds to π_v having the form

$$\pi_v = \frac{\exp\left(\beta_0 + \sum_{j=1}^{p} \beta_j z_{jv}\right)}{1 + \exp\left(\beta_0 + \sum_{j=1}^{p} \beta_j z_{jv}\right)} . \tag{7.2}$$

The likelihood function for the logistic regression model is given by

$$L = \prod_{v=1}^{N} \pi_v^{y_v}(1-\pi_v)^{1-y_v} ,$$

since case v contributes with the factor π_v if $y_v = 1$ and $1 - \pi_v$ if $y_v = 0$. This means that the log-likelihood function takes the form

$$\ln L = \sum_{v=1}^{N} y_v \cdot \ln\left(\frac{\pi_v}{1-\pi_v}\right) + \sum_{v=1}^{N} \ln(1-\pi_v) . \tag{7.3}$$

If we insert (7.1) in this expression, we get

$$\ln L = \beta_0 \sum_{v=1}^{N} y_v + \sum_{j=1}^{p} \beta_j \sum_{v=1}^{N} y_v z_{jv} + \sum_{v=1}^{N} \ln(1-\pi_v) . \tag{7.4}$$

Equation (7.4) shows that the model belongs to an exponential family and that the sufficient statistics are

$$t_0 = \sum_{v=1}^{N} y_v = y.$$

and

$$t_j = \sum_{v=1}^{N} y_v z_{jv}, \quad j=1,\ldots,p.$$

7.2 Estimation in the logistic regression model

Since $E[Y_v] = \pi_v$ when the possible values of Y_v are 1 and 0, it follows from Theorem 2.1, that the likelihood equations become

$$t_0 = y. = \sum_{v=1}^{N} \pi_v \qquad (7.5)$$

and

$$t_j = \sum_{v=1}^{N} y_v z_{jv} = \sum_{v=1}^{N} \pi_v z_{jv}, \quad j=1,\ldots,p, \qquad (7.6)$$

and that equations (7.5) and (7.6) have a unique set of solutions if $(t_0, t_1, \ldots, t_p)$ is an interior point in the convex extension of the support. For logistic regression the support is relatively simple and there are accordingly simple rules for when the likelihood equations have a unique set of solutions.

Note: Strictly speaking we can not apply the theorems in section 2.3 to the logistic regression model. The reason is that the Y's are not identically distributed. The theorems are basically valid, however, because the likelihood function, expressed by means of the canonical parameters and the sufficient statistics has the form (2.7), where for non-identical distributed variables the term $nK(\tau)$ has the slightly more complicated form

$$\Sigma_i K_i(\tau).$$

Thus for the logistic regression model

$$\Sigma_v K_v(\beta) = -\Sigma_v \ln(1 - \pi_v)$$

where π_v is a function of the vector $\beta = (\beta_0, \beta_1, \ldots, \beta_p)$.

To illustrate the form the support takes for a logistic regression model, consider the case $p = 1$, where there are two parameters β_0 and β_1 in the model and the sufficient statistics are

$$t_0 = y. = \sum_{v=1}^{N} y_v \tag{7.7}$$

and

$$t_1 = \sum_{v=1}^{N} y_v z_{v1} . \tag{7.8}$$

Since the possible values of y_v are 1 and 0, the possible values of t_0 are all integers between 0 and N. Now let

$$z_{(1)} \leq ... \leq z_{(N)} ,$$

be the ordered values of z_{v1} and let $y. = t_0$. Then the limits for t_1 are the sum of the t_0 smallest and the sum of the t_0 largest z-values, i.e.

$$z_{(1)} + ... + z_{(t_0)} \leq t_1 \leq z_{(N-t_0+1)} + ... + z_{(N)} , \tag{7.9}$$

as can be verified directly from (7.8). The inequalities (7.9) together with

$$0 \leq t_0 \leq N \tag{7.10}$$

thus form the limits for the support. The convex extension is then constructed simply by plotting the limits (7.9) in a (t_0, t_1)-plane for each value of t_0 between 0 and N and connecting the plotted points by straight lines. For all observed cases $y_1,...,y_N$, where (t_0, t_1) given by (7.7) and (7.8) are inside these lines, there is a unique solution to the likelihood equations (7.5) and (7.6) for p = 1. As an example, let N = 12 and the z-values be

$$(z_{11}, ... ,z_{1.12}) = (-5, 8, 11, 2, 0, 1, 5, 5, -5, -3, 0, 3) .$$

The convex extension of the support for these values is shown in Figure 7.1.

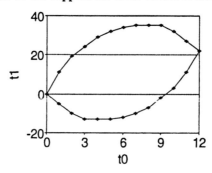

FIGURE 7.1. The convex extension of the support for a logistic regression model with p=1.

LOGISTIC REGRESSION ANALYSIS

For p > 2 the convex area, within the interior of which the likelihood equations have unique solutions, becomes more complicated. However in 1984 Albert and Andersson (cf. the bibliographic notes) introduced an algebraic criterion which makes it easy to check whether there are unique solutions to the likelihood equations. This criterion is based on the following definition:

DEFINITION: *A set of observations $y_1, \ldots, y_N$ are said to be **quasi-complete separated** if there exist constants $(\alpha_0, \ldots, \alpha_p)$, with at least one $\alpha_j \neq 0$, $j = 1, \ldots, p$, such that*

$$\alpha_0 + \alpha_1 z_{1v} + \ldots + \alpha_p z_{pv} \geq 0$$

for all $y_v = 1$, and

$$\alpha_0 + \alpha_1 z_{1v} + \ldots + \alpha_p z_{pv} \leq 0$$

for all $y_v = 0$.

Quasi-complete separation means that there is a plane, spanned by the vector $(\alpha_0, \ldots, \alpha_p)$, in the (p + 1)-dimensional Euclidian space, such that all z's corresponding to $y_v = 1$ are situated on one side of (for example above) this plane, while all z's corresponding to $y_v = 0$ are situated on the opposite side of the plane. For points on the plane itself, that is points for which

$$\alpha_0 + \alpha_1 z_{1v} + \ldots + \alpha_p z_{pv} = 0$$

it does not matter whether y_v has the value 1 or the value 0.

For p=1 the criterion is that

$$\alpha_0 + \alpha_1 z_{1v} \geq 0 \text{ for } y_v = 1$$

and

$$\alpha_0 + \alpha_1 z_{1v} \leq 0 \text{ for } y_v = 0 .$$

If $\alpha_1 > 0$ this leads to

$$z_{1v} \geq -\alpha_0/\alpha_1 \text{ for } y_v = 1$$

and

$$z_{1v} \leq -\alpha_0/\alpha_1 \text{ for } y_v = 0 ,$$

while we for $\alpha_1 < 0$ get the opposite inequalities. Hence for p = 1 the observations are quasi-complete separated if there is a constant c, such that $z_{1v} \geq c$ for all observations with $y_v = 1$ and $z_{1v} \leq c$ for all observations with $y_v = 0$, or vice versa $z_{1v} \leq c$ for all $y_v = 1$ and $z_{1v} \geq c$ for all $y_v = 0$.

For p = 2 the observations are quasi-complete separated if there is a line

$$z_{2v} = c_0 + c_1 z_{1v} ,$$

such that all observations with $y_v = 0$ are either on the line or on the same side of the line, while all observations with $y_v = 1$ are either on the line or on the opposite side of the line.

In summary we have:

THEOREM 7.1: *There exists a unique set of solutions to the likelihood equations (7.5), (7.6) except if the observations are quasi-complete separated*

7.3 Numerical solution of the likelihood equations

In order to solve the likelihood equations numerically the so-called **Newton-Raphson method** is used. Briefly, the method is based on the following arguments.

If we introduce the additional z-values $z_{0v} = 1$ and let Z be the matrix of dimension $(p+1) \times N$ with elements z_{jv}, $j = 0,...,p$, $v = 1,...,N$, the likelihood equations (7.5) and (7.6) can be written in the matrix form

$$yZ' - \pi Z' = 0, \qquad (7.11)$$

where $y = (y_1, ..., y_N)$, $\pi = (\pi_1, ..., \pi_N)$ and 0 is a vector of p+1 zeros. If we perform a Taylor expansion on (7.11) and include only one term, we get with $\beta = (\beta_0,...,\beta_p)$, β^0 the point of expansion and π^0 the vector of π's computed with $\beta = \beta^0$

$$0 \sim yZ' - \pi^0 Z' + (\beta - \beta^0)D^0 \qquad (7.12)$$

since β is the vector of ML-estimates and therefore satisfies (7.11). The matrix D^0 is of dimension $(p+1) \times (p+1)$ and contains the derivatives

$$\frac{\partial (\sum_v \pi_v z_{jv})}{\partial \beta_q}, \; j,q = 0,...p \; .$$

If now β^0 is an initial guess for the vector of β's, the Newton-Raphson method is to solve (7.12) with respect to β in order to get new improved estimates for the β's. The solution is

$$\beta^1 = \beta^0 - (yZ' - \pi^0 Z')(D^0)^{-1}, \qquad (7.13)$$

where β^1 is the improved estimate. The improvement of β is repeated until it does not (within a specified limit) change any more.

In order to apply the algorithm (7.13) we need a simple expression for the matrix D^0 of derivatives. These derivatives have, however, a very simple form. To see this we recall from (7.2) that the response probability π_v is given by

$$\pi_v = \frac{\exp(\beta_0 + \sum_{j=1}^{p} \beta_j z_{jv})}{1 + \exp(\beta_0 + \sum_{j=1}^{p} \beta_j z_{jv})}.$$

Differentiation with respect to z_{jv} then yields

$$\frac{\partial \pi_v}{\partial \beta_q} = \frac{z_{qv} \exp(\beta_0 + \sum_{j=1}^{p} \beta_j z_{jv})}{\left(1 + \exp(\beta_0 + \sum_{j=1}^{p} \beta_j z_{jv})\right)^2} = z_{qv} \pi_v (1 - \pi_v) ,$$

so that

$$\frac{\partial (\sum_v \pi_v z_{jv})}{\partial \beta_q} = \sum_v z_{jv} z_{qv} \pi_v (1 - \pi_v) .$$

The matrix $\mathbf{D}^0$, therefore, has the matrix form

$$\mathbf{D}^0 = \mathbf{ZWZ}' ,$$

where $\mathbf{W}$ is a diagonal matrix of dimension N with diagonal elements $\pi_v(1 - \pi_v)$. According to the Newton-Raphson method we thus improve a set of initial estimates $\beta^0 = (\beta_0^0, \ldots, \beta_p^0)$ as

$$\beta^1 = \beta^0 + (\mathbf{y}\mathbf{Z}' - \pi^0 \mathbf{Z}')(\mathbf{Z}\mathbf{W}^0 \mathbf{Z}')^{-1} , \quad (7.14)$$

where $\pi^0 = (\pi_1^0, \ldots, \pi_N^0)$, $\mathbf{W}^0$ is diagonal with diagonal elements $\pi_v^0(1 - \pi_v^0)$ and π_v^0 is (7.2) with the β's replaced by $\beta_0^0, \ldots, \beta_p^0$.

The algorithm (7.14) is computational very simple and effective. In addition it can be shown that

$$\text{var}[\beta] = (\mathbf{ZWZ}')^{-1} . \quad (7.15)$$

Estimates for the standard errors of the β's are thus a biproduct of the estimation procedure.

7.4 Checking the fit of the model

The logistic regression model can be checked in two different ways. We can check the model fit by graphical methods, primarily by using residual diagrams, or we can use a goodness of fit test statistic. For the goodness of fit test we need more than one observed y value for each distinct combination of values of the explanatory variables, or at least that we for a relatively large and representative percentage of distinct values of the explanatory variables have more than one observed y-value.

It is, therefore, convenient to distinguish between two schemes A and B:

Scheme A: There is one observed y-value for each distinct combination of the explanatory variables.

Scheme B: There are several observed y-values for each distinct combination of the explanatory variables.

It is not a question of a mathematically or logically sharp distinction between two schemes, but more a question of one for the applications' practical distinction. In some situations we will have a mixture of the two schemes.

For scheme B we have to change the notations slightly. Thus let $i = 1,...,I$ be an index, running over all distinct combinations of the explanatory variables, such that

$$z_i = (z_{1i}, ... , z_{pi})$$

is a typical combination of observed explanatory variables. There are thus I different vectors z_i and we introduce the counts

n_i = number of cases y_v with values $z_{1i},...,z_{pi}$ of the explanatory variables.

and

x_i = number of these n_i cases for which $y_v = 1$.

This implies that the N cases y_v can be divided into I **groups** of sizes $n_1, n_2, ... ,n_I$, where of course

$$n_1 + n_2 + ... + n_I = N .$$

All cases in a group have the same values of the z's, while at least one z_{jv} is different if two observations belong to different groups. It further follows that π_v given by (7.2) is constant in group i and has the value

$$\pi_i = \frac{\exp\left(\beta_0 + \sum_j \beta_j z_{ji}\right)}{1 + \exp\left(\beta_0 + \sum_j \beta_j z_{ji}\right)} . \tag{7.16}$$

Since the random variables corresponding to the y's are independent, the same is true for the random variables $X_1,...,X_I$ corresponding to the x's. Finally it is easy to see that for each i, X_i has the binomial distribution

LOGISTIC REGRESSION ANALYSIS

$$X_i \sim \text{bin}(n_i, \pi_i), \quad (7.17)$$

since the probability of observing π_i for $Y_v = 1$ is constant within the i'th group.

If in (7.5) and (7.6) we first sum within the groups and then over the groups, noting that z_{jv} is constant and equal to z_{ji} within a group, the likelihood equations (7.5) and (7.6) expressed in terms of the x's become

$$\sum_i x_i = \sum_i n_i \pi_i \quad (7.18)$$

and

$$\sum_i x_i z_{ji} = \sum_i n_i \pi_i z_{ji}, \quad j = 1, \ldots, p \quad (7.19)$$

It follows from (7.18) and (7.19) that the likelihood equations only depend on the x's thus being the sufficient statistics. The likelihood function can accordingly be formulated based on the binomial distributions (7.17) of the X's. From independence and (7.17) it follows that the likelihood function is

$$L = \prod_{i=1}^{I} \binom{n_i}{x_i} \pi_i^{x_i} (1 - \pi_i)^{n_i - x_i}, \quad (7.20)$$

where π_i is given by (7.16).

The likelihood function (7.20) shows that there is an alternative to the logistic regression model, namely a model with no restrictions on the binomial parameters π_i. If we maximize (7.20) without restrictions on the π's, the ML-estimates become

$$\hat{\pi}_i = \frac{x_i}{n_i}.$$

It is, therefore, possible to check whether a logistic regression model fits the data by the likelihood ratio test

$$\frac{L(\hat{\pi}_1, \ldots, \hat{\pi}_I)}{L(\tilde{\pi}_1, \ldots, \tilde{\pi}_I)},$$

where the likelihood function in the numerator is maximized under the logistic regression model with

$$\hat{\pi}_i = \frac{\exp(\beta_0 + \sum_j \beta_j z_{ji})}{1 + \exp(\beta_0 + \sum_j \beta_j z_{ji})} \quad (7.21)$$

and the likelihood function in the denominator is maximized by $\tilde{\pi}_i = x_i/n_i$.

According to Theorem 2.5 the likelihood ratio (transformed by -2 ln) is approximately χ^2-distributed. The hypothesis that a logistic regression model fits the data, rather than a product of unrestricted binomial distributions, can thus be tested by the test statistic

$$Z = -2\ln L(\hat{\pi}_1,\ldots,\hat{\pi}_I) + 2\ln L(\tilde{\pi}_1,\ldots,\tilde{\pi}_I) \qquad (7.22)$$

The larger the observed value of (7.22), the larger is it possible to make the likelihood by maximizing without restrictions rather than by maximizing under the restriction of a logistic regression model. Hence we should reject the hypothesis that a logistic regression model fits the data if the observed value of Z is large.

According to Theorem 2.5 the number of degrees of freedom for the approximating χ^2-distribution is the number of constrained parameters under the hypothesis. In the unconstrained model there are I canonical parameters, namely one probability parameter for each binomial distribution. In the logistic regression model there are p+1 canonical parameters. Hence the number of constrained parameters are

$$df = I - p - 1 .$$

We accordingly reject a logistic regression model as a satisfactory fit to the data, if the level of significance computed approximately as

$$p = P(Q \geq z) , \qquad (7.23)$$

where $Q \sim \chi^2(I - p - 1)$, is smaller than a certain critical level.

Note: When we counted the degrees of freedom for (7.22), we counted differences between canonical parameters. One might object that the π's and the β's are not linearly connected. But in fact the canonical parameter in the i'th binomial distribution is

$$\tau_i = \ln\{\pi_i/(1 - \pi_i)\}$$

and τ_i has under the logistic regression model the form

$$\tau_i = \beta_0 + \Sigma_j \beta_j z_{ji} , \qquad (7.24)$$

such that the I - p - 1 constrained canonical parameters are in fact linear constraints on the canonical parameters of the larger model.

Note also that (7.24) has the matrix form

$$\tau = \mathbf{Z}'\beta$$

with solution

$$\beta = (\mathbf{ZZ}')^{-1}\mathbf{Z}\tau .$$

It is thus a condition for Z to have I - p - 1 degrees of freedom that **Z** and therefore **ZZ**' has rank p + 1.

As usual the χ^2-approximation requires that the expected numbers are not too small. It would be wise for example to only use the χ^2-approximation if both $n_i\hat{\pi}_i > 3$ and $n_i\hat{\pi}_i > 3$.

Since the model is not a single multinomial distribution, we can not use the results from section 2.9 concerning the distribution of the residuals directly. It can be shown, however, (see biographical notes) that for a logistic regression model the variance of the residuals can be approximated by

$$\text{var}\left[X_i - n_i\hat{\pi}_i\right] = n_i\pi_i(1-\pi_i)(1-h_i) , \qquad (7.25)$$

where h_i is the diagonal element in the matrix

$$W^{\frac{1}{2}}Z'(ZWZ')ZW^{\frac{1}{2}}$$

and **W** and **Z** are the matrices, introduced in section 7.3, except that for case B **W** must be redefined as a diagonal matrix of dimension I with elements $n_i\pi_i(1 - \pi_i)$ and **Z** must be redefined as a matrix of dimension $(p + 1) \times I$ and elements z_{ji} with $z_{0i} = 1$.

Equation (7.25) implies that **standardized residuals** can be defined as

$$r_i = \frac{x_i - n_i\hat{\pi}_i}{\sqrt{n_i\hat{\pi}_i(1-\hat{\pi}_i)(1-\hat{h}_i)}} , \qquad (7.26)$$

where $\hat{h}_i$ is obtained by replacing π_i with $\hat{\pi}_i$ in **W**. The residuals (7.26) are approximately normally distributed with mean 0 and variance 1.

The model fit can be checked graphically by plotting the residuals (7.26) against the expected values $n_i\hat{\pi}_i$ or against each of the explanatory variables. Such **residual diagrams** can point out model deviations. From the fact that the residuals (7.26) are standardized it follows that residuals which are numerically larger than 2, call for further inspection. We postpone demonstrations of the use of the standardized residuals (7.62) to section 7.6.

EXAMPLE 7.1. *Stress at work.*
From the data base of the Danish Welfare Study 1976 it is possible to extract information on the persons in the sample, who felt that they suffered from stress at work. The response variable thus has value 1 for a person feeling stress at work and the value 0 if he or she did not suffer from stress. We shall try to describe the variation in the stress variable by the following explanatory variables:

Age, divided into five age intervals, which we score by the interval midpoints

20-29 year: 25
30-39 year: 35
40-49 year: 45
50-59 year: 55
60-69 year: 65

Sector, with categories Private, scored 1 and Public, scored -1.
Employment, with categories White collar, scored 1 and Blue collar scored -1.

Table 7.1 shows the number of persons in the sample feeling stress at work for each combination of the explanatory variables.

TABLE 7.1. The number of persons in the Welfare Study feeling stress at work for each combination of Sector, Employment and Age.

	Response		Explanatory variables		
Case	Number with stress x_i	Number of persons n_i	Sector z_{1i}	Employment z_{2i}	Age z_{3i}
i=1	39	265	1	1	25
2	63	253	1	1	35
3	38	155	1	1	45
4	23	111	1	1	55
5	4	45	1	1	65
6	46	298	1	-1	25
7	65	292	1	-1	35
8	46	211	1	-1	45
9	35	188	1	-1	55
10	13	82	1	-1	65
11	30	189	-1	1	25
12	41	260	-1	1	35
13	19	137	-1	1	45
14	19	118	-1	1	55
15	7	44	-1	1	65
16	8	55	-1	-1	25
17	7	58	-1	-1	35
18	12	60	-1	-1	45
19	12	73	-1	-1	55
20	2	33	-1	-1	65
Totals	529	2927	215	37	20995

Source: The data base of the Danish Welfare Study. Cf. Example 3.2.

The last line in Table 7.1 gives the totals

LOGISTIC REGRESSION ANALYSIS

$$x_. = \sum_i x_i ,$$

$$n = \sum_i n_i$$

and the sufficient statistics

$$t_j = \sum_i x_i z_{ji} , \; j=1,\ldots,3 .$$

In this example, we can use the Albert and Andersson criterion to verify that there is in fact a unique solution to the likelihood equations. In Figure 7.2 we have plotted all combinations of values (z_1, z_2, z_3) represented in Table 7.1. Since all x_i's satisfy

$$0 < x_i < n_i ,$$

the points in the 3-dimensional space, shown in Figure 7.2, are quasi-complete separated only if there is a plane on which all the points are located. Since this is obviously not the case we can conclude that there is, in fact, a unique set of solutions to the likelihood equations.

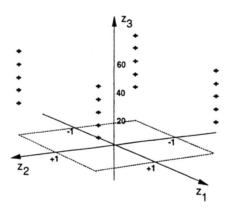

FIGURE 7.2. Configuration of (z_1, z_2, z_3)-points for the data in Table 7.1.

The estimated parameters with standard errors are shown in Table 7.2.

TABLE 7.2. Estimated regression parameters for the data in Table 7.1 with standard errors.

Variable	Parameter	Estimate	Standard error
Intercept	β_0	-1.629	0.165
Sector	β_1	0.161	0.054
Employment	β_2	0.034	0.050
Age	β_3	0.002	0.004

The test statistic (7.22) has observed value

$$z = 25.15 \, , \, df = 16 \, ,$$

with level of significance p = 0.067, indicating a satisfactory fit by a logistic regressions model.

EXAMPLE 7.2. *Indoor climate.*
The Danish Institute for Building Research in 1970 carried out an investigation of the indoor climate in Danish schools. The students in 3 school classes was asked at different occasions whether they felt that the indoor climate was pleasant or not so pleasant. The response 1 indicated that the indoor climate was pleasant, while response 0 indicated a not so pleasant indoor climate. Simultaneously the Building Research Institute measured 6 indicators of the actual (or "objective") indoor climate for each occasion in each class room. These indicators were as follows.

T: *Temperature.*
M: *The degree of moisture.*
C: *The amount of carbon dioxide in the air.*
F: *The amount of fresh air.*
D: *The amount of dust.*
V: *The degree of ventilation.*

Table 7.3 shows the collected data, i.e. for each class on each occasion the number of students claiming a pleasant indoor climate together with the class size and the 6 measurements of the "objective" indoor climate indicators.

LOGISTIC REGRESSION ANALYSIS 171

TABLE 7.3 The number of students (x) claiming a pleasant indoor climate shown together with class size and 6 objective indicators for 3 Danish school classes on different days.

					Objective indicators				
Class	Date	Class size	x	T	M	C	F	D	V
7A	4/3	19	18	22.0	30	0.09	8.5	0.20	230
7A	3/3	20	16	21.5	25	0.11	6.1	0.08	230
7A	2/3	19	4	21.5	25	0.11	4.8	0.06	230
8A	4/3	18	13	18.5	25	0.09	9.2	0.07	236
8A	3/3	14	12	20.0	25	0.05	8.7	0.08	236
8A	2/3	18	4	20.0	25	0.11	5.2	0.12	236
8A	18/3	17	14	20.5	30	0.08	13.1	0.09	249
8A	17/3	19	18	21.0	30	0.08	12.5	0.06	249
9B	18/3	16	9	21.5	30	0.09	8.7	0.07	215
9B	17/3	18	8	21.0	30	0.07	9.3	0.09	215

Source: Unpublished data from the Danish Institute for Building Research.

We shall now analyze the data in Table 7.3 using an analysis based on a logistic regression model. First, a logistic regression model is checked by the Z-test statistic (7.22). The observed value of Z is
$$z = 176.05 - 169.58 = 6.47, df = 3,$$

with level of significance p = 0.091 indicating a satisfactory fit. The number of degrees of freedom 3 are obtained as the difference between I = 10 and p+1 = 7.

Table 7.4 shows the ML-estimates of the regression parameters and their standard errors.

TABLE 7.4. ML-estimates for the regression parameters with standard errors for the data in Table 7.3.

Variable	Parameter	Estimate	Standard error
Intercept	β_0	4.656	9.064
T: Temperature	β_1	1.320	0.342
M: Moisture	β_2	-1.141	0.290
C: Carbon dioxide	β_3	20.290	17.700
F: Fresh air	β_4	1.449	0.330
D: Dust	β_5	25.305	7.560
V: Ventilation	β_6	-0.070	0.036

The estimates seem to indicate that only the explanatory variables Temperature, Moisture, Fresh air and Dust contribute significantly to describe the variation in the response variable: Pleasant or unpleasant indoor climate.

7.5 Hypothesis testing

In logistic regression analysis there are two main types of hypotheses to be tested:

$$H_j: \beta_j = 0, \quad j = 1,...,p$$

and

$$H_{(j)}: \beta_j = ... = \beta_p = 0, \quad j = 1,...,p.$$

$H_{(j)}$ is mainly used if we have reasons a priori to believe that none of the explanatory variables connected with $\beta_j, ... ,\beta_p$ contribute significantly to the regression model. The connection between $H_{(j)}$ and H_j is that $H_{(j)}$ is true if and only if all the hypotheses $H_j, H_{j+1}, ... ,H_p$ are true.

Consider first the test for H_j. For this test we need the ML-estimates for the β's under the hypothesis. Let us for convenience start with j = p, where $H_p = H_{(p)}$. If $L(\beta_0, ... ,\beta_p)$ is the likelihood function (7.20) with the π's given by (7.16) the ML-estimates $(\tilde{\beta}_0, ... ,\tilde{\beta}_{p-1})$ under H_p are defined by the relationship

$$L(\tilde{\beta}_0,...,\tilde{\beta}_{p-1},0) = \max \{ L(\beta_0,...,\beta_{p-1},0) \}.$$

The arguments leading to the Z-test statistic (7.22) then yield that H_p can be tested by the test statistic

$$Z(H_p) = -2\ln L(\tilde{\beta}_0,...,\tilde{\beta}_{p-1},0) + 2\ln(\hat{\beta}_0,...,\hat{\beta}_{p-1},\hat{\beta}_p), \quad (7.27)$$

where $(\hat{\beta}_0, ... ,\hat{\beta}_p)$ are the ML-estimates for the regression parameters in the full

LOGISTIC REGRESSION ANALYSIS

model. Since $\beta_p = 0$ is the only canonical parameter, which is specified under H_p, (7.27) is asymptotically χ^2-distributed with 1 degree of freedom. Hence the hypothesis $H_p : \beta_p = 0$ can be tested by evaluating the level of significance p as

$$p = P(Q \geq z(H_p)), \qquad (7.28)$$

where $z(H_p)$ is the observed value of $z(H_p)$ and $Q \sim \chi^2(1)$. If p is less than a certain critical value, we reject H_p. How to choose this critical value depends on the circumstances.

Note: The letter p is used both for a level of significance and for the number of explanatory variables. The meaning of p should always be clear from the context.

If H_p is accepted, we say that the p'th explanatory variable does not contribute to the description of the variation in the response variable. This is a neutral statement. A statement like " the p'th explanatory variable does not influence the response variable" is more suggestive. It would for example implicate, or at least hint at, a causal relationship. It also suggests that explanatory variable p has an effect on the response variable, independently of how the other explanatory variables vary.

In a logistic regression analysis there are p explanatory variables to consider. Hence it is often practical to organize the analysis such that the contributions of the explanatory variables are evaluated in a certain order. One often used possibility is to take the variables in an order determined by the magnitudes of the significance levels

$$p = P(Q \geq z(H_j)),$$

where $z(H_j)$ is the observed value of

$$Z(H_j) = 2\ln L(\tilde{\beta}_0,...,\tilde{\beta}_{j-1},0,\tilde{\beta}_{j+1},...,\tilde{\beta}_p) + 2\ln(\hat{\beta}_0,...,\hat{\beta}_{p-1},\hat{\beta}_p) \qquad (7.29)$$

Here, as before, the $\tilde{\beta}$'s are the ML-estimates under H_j and the $\hat{\beta}$'s are the ML-estimates in the unrestricted model. The procedure is then first to consider the hypothesis with the highest level of significance, then the hypothesis with the second highest level of significance and so on.

If possible one should, however, discusses the contributions of the explanatory variables in an order which reflect a priori considerations on the part of the principal investigator who originally collected the data.

The hypotheses $H_{(j)}$ are the appropriate instruments for a **sequential testing** of the contributions from the explanatory variables. Since $H_{(j)}$ is the hypothesis that $p - j + 1$ of the explanatory variables can be omitted, or equivalently that it suffices to include $j - 1$ explanatory variables in the logistic regression model, the test statistic for $H_{(j)}$ is

$$Z(H_{(j)}) = -2\ln L(\tilde{\beta}_0,...,\tilde{\beta}_{j-1},0,...,0) + 2\ln(\hat{\beta}_0,...,\hat{\beta}_{p-1},\hat{\beta}_p), \quad (7.30)$$

where $\hat{\beta}_0,...,\hat{\beta}_{j-1}$ are the ML-estimators under $H_{(j)}$.

The test statistic (7.30) is asymptotically χ^2-distributed with p+1-j degrees of freedom, since p+1-j β's are specified under $H_{(j)}$. We, therefore, reject the hypothesis if the level of significance is below a certain critical level
$$p = P(Q \geq z(H_{(j)})) ,$$
Here $z(H_{(j)})$ is the observed value of (7.30), and $Q \sim \chi^2(p+1-j)$.

The tests for the sequential procedure now become

$$Z(H_{(j)}|H_{(j+1)}) = -2 \ln L(\hat{\beta}_0, ... ,\hat{\beta}_{j-1},0,...,0) + 2 \ln L(\hat{\beta}_0,...,\hat{\beta}_j,0,...,0) , \quad (7.31)$$

where $\hat{\beta}_0,...,\hat{\beta}_{j-1}$ are the ML-estimates under $H_{(j)}$ and $\hat{\beta}_0,...,\hat{\beta}_j$ the ML-estimates under $H_{(j+1)}$. It follows from (7.30) and (7.31), that

$$Z(H_{(j)}|H_{(j+1)}) = Z(H_{(j)}) - Z(H_{j+1}) .$$

Since exactly one canonical parameter, β_j, is specified under $H_{(j)}$ with $H_{(j+1)}$ as alternative, it further follows that (7.31) is approximately χ^2-distributed with 1 degree of freedom.

EXAMPLE 7.2 (continued). *The estimates and their standard errors in Table 7.4 suggested that the explanatory variables C and V did not contribute to the description of the variation in the response variable. Table 7.5 shows both the observed values of the test statistics $Z(H_j)$ and of the sequential test statistics $Z(H_{(j)} | Z(H_{(j+1)}))$ with the associated degrees of freedom and levels of significance. In Table 7.5 the system of excluding the explanatory variable with the highest significance level for $Z(H_{(j)})$ first, then the variable with the second highest level of significance for $Z(H_{(j)})$, and so on, is applied.*

TABLE 7.5. Direct and sequential tests with degrees of freedom (df) and levels of significance (p) for the indoor climate data.

Hypothesis	Variable omitted	Parameters equal to 0	$z(H_{(j)})$	df	p	$z(H_{(j)}|H_{(j+1)})$	df	p
$H_{(6)}$	C	$\beta_3=0$	1.25	1	0.264	1.25	1	0.264
$H_{(5)}$	V	$\beta_3=\beta_6=0$	3.82	2	0.148	2.58	1	0.108
$H_{(4)}$	D	$\beta_3=\beta_6=\beta_5=0$	14.67	3	0.002	10.85	1	0.001

Note 1: The notation $H_{(6)}$, $H_{(5)}$ and so on, does not follow the notation in the text, since explanatory variable 3 is omitted first, not the last number 6, etc. But the system should be clear enough from the entries in the table.

Note 2: $H_{(6)}$ is tested against the full regression model in column 7.

The significance levels in Table 7.5 reveal that we were right that only the variables Carbondioxide and Ventilation can be omitted from the model.

EXAMPLE 7.1 (continued). *The table corresponding to Table 7.5 for the stress at work data is shown in Table 7.6. In this case the order of the hypotheses $H_{(3)}$, $H_{(2)}$ and $H_{(1)}$ corresponds to the subscripts of the β's.*

TABLE 7.6. Direct and sequential tests with degrees of freedom (df) and levels of significance (p) for the stress at work data.

Hypothesis	Variable omitted	Parameters equal to 0	$z(H_{(j)})$	df	p	$z(H_{(j)}|H_{(j+1)})$	df	p
$H_{(3)}$	Age	$\beta_3=0$	0.15	1	0.695	0.15	1	0.695
$H_{(2)}$	Empl.	$\beta_3=\beta_2=0$	0.56	2	0.776	0.40	1	0.526
$H_{(1)}$	Sector	$\beta_3=\beta_2=\beta_1=0$	9.02	3	0.029	8.47	1	0.004

Note: $H_{(3)}$ is tested against the full regression model in column 7.

Both Age and Employment can thus be left out of the model without making the fit significantly worse. The explanatory variable Sector on the other hand contributes significantly to the description of the response variable. The sign of β_1 in Table 7.2 shows that persons with $z_{1i} = 1$ more often than persons with $z_{1i} = -1$ are stressed at work. Since persons employed in the private sector are scored 1 and persons employed in the public sector are scored -1, this means, as expected, that persons in the private sector are stressed more often than persons in the public sector.

In scheme A, where it is not possible to check the logistic regression model by a goodness of fit test, we can still estimate the regression parameters and evaluate their contributions to describing the variation in the response variable. This means that we can test the contribution of variable z_j by the test statistic (7.29), or sequentially by the test statistic (7.31). Both test statistics are differences between

$$-2\ln L = 2\ln(\beta_0,\ldots,\beta_p) ,$$

or

$$-2\ln L = 2\ln(\beta_0,\ldots,\beta_j,0,\ldots,0) ,$$

depending on which explanatory variables are included in the model. Hence in order to test hypotheses concerning the contributions of the explanatory variables all we need is a table of -2lnL and the corresponding number of degrees of freedom for various sets of the explanatory variables included in the model. From this table we can then compute the desired z-test statistics by subtracting values of -2lnL. Most packages provide the values of -2lnL for specified models.

EXAMPLE 7.3. *Public and private employment.*
From the data base of the Danish Welfare Study we have for this example extracted the following variables for further analysis.

> A: *Employment sector, with categories Public sector and private sector.*
> B: *Social group, with the usual 4 groups.*
> C: *Sex, with categories Male and Female.*
> D: *Income in Danish kroner, grouped in income intervals.*
> E: *Age in years.*
> F: *Urbanization, with categories Copenhagen, Suburbs of Copenhagen, Three biggest cities (outside Copenhagen), Other cities and Countryside.*

Here Employment sector is the response variable and we want to analyze which of the explanatory variables contribute to explain the variation in the response variable.

The complete list of -2lnL for different sets of explanatory variables included in the model is shown in Table 7.7.

TABLE 7.7 The complete list of -2lnL for different sets of explanatory variables included.

Explanatory variables included	-2 ln L	Explanatory variables included	-2 ln L
B C D E F	3305.21	B C	3333.15
B C D E	3308.62	B D	3501.40
B C D F	3327.69	B E	3501.26
B C E F	3305.22	B F	3503.93
B D E F	3471.80	C D	3495.64
C D E F	3473.82	C E	3514.48
B C D	3332.60	C F	3508.09
B C E	3308.66	D E	3656.52
B C F	3327.97	D F	3638.67
B D E	3482.39	E F	3633.42
B D F	3488.51	B	3514.93
B E F	3492.10	C	3527.26
C D E	3485.37	D	3663.40
C D F	3482.80	E	3657.59
C E F	3497.30	F	3665.18
D E F	3633.16	None	3665.04

LOGISTIC REGRESSION ANALYSIS   177

From Table 7.7 we can derive a sequence of test statistics of the form (7.31) for successive omissions of explanatory variables. This is shown in Table 7.8.

TABLE 7.8 A sequence of test statistics z(H|H*) for possible exclusion of explanatory variables with degrees of freedom and levels of significance.

Hypothesis	Variables included	Variable omitted	z(H|H.)	df	Level of significance
H_0	B C D E F	-	-	-	-
H_1	B C E F	D	0.01	1	0.920
H_2	B C E	F	3.44	1	0.064
H_3	B C	E	24.49	1	0.000
H_4	B	C	181.78	1	0.000
H_5	None	B	150.11	1	0.000

Table 7.8 thus shows that Sex and Urbanization contribute to explain the variation in the response variable, while Social group, Income and Age are explanatory variables which are of importance if we want to predict whether a given person is in the private or in the public sector.

7.6 Diagnostics

The main diagnostics for detection of model deviations are the standardized residuals (7.26). In recent years a number of additional residuals have been suggested. One example are the individual terms in the Z-test statistic

$$-2 \ln (x_i) + 2 \ln (n_i \hat{\pi}_i),$$

called **deviances**. Scaled by their standard errors to be approximately normally distributed with mean 0 and variance 1, they are used in the same way as standardized residuals. Another type of diagnostic is **Cook's distance**, which is a measure of the influence of each particular group on the estimates of β's. Cook's distance for group i is defined as

$$C_i = \frac{1}{p+1}(\beta - \beta^{(i)})(Z\hat{W}Z')(\beta - \beta^{(i)})',$$

where $\hat{\beta}$ is the vector of ML-estimates and $\hat{\beta}^{(i)}$ a vector, which contains the ML-estimates one would obtain if the i'th group were omitted from the data. The factor $(Z\hat{W}Z')$ is a scaling factor since it follows from (7.15) that

$$\hat{var}[\hat{\beta}] = (Z\hat{W}Z')^{-1}.$$

The so-called hat matrix plays an important role both for standardized residuals and

Cook's distances. The hat matrix is defined by

$$\hat{H} = \hat{W}^{1/2} Z'(Z\hat{W}Z')Z\hat{W}^{1/2} ,\qquad(7.32)$$

where W is the diagonal matrix, which was introduced in connection with (7.25) and (7.26), but now with the π_i's replaced by their ML-estimates, i.e. $\hat{W}$ has diagonal elements $n_i \hat{\pi}_i (1 - \hat{\pi}_i)$. The diagonal elements $\hat{h}_i$ of the hat matrix are called **leverages**. Note that there is a leverage for each group i.

Equation (7.26) shows that we need the elements of the hat matrix in order to compute the standardized residuals. In addition it can be shown (see biographical notes) that Cook's distance for logistic regression can be approximated by

$$C_i = \frac{1}{p+1} r_i^2 \frac{\hat{h}_i}{1-\hat{h}_i} .\qquad(7.33)$$

showing that Cook's distance is in general large if the standardized residual r_i is large, but also if the leverage $\hat{h}_i$ is close to its maximal value 1. Leverages close to one are more common in logistic regression models with several responses for each distinct combination of the explanatory variables (scheme B), than in scheme A, or in ordinary regression analysis. It is in particular when n_i is large, that we risk large leverages. This is because groups based on many observations potentially contribute more to the estimates than if the number of cases n_i in a group is small.

Cook's distance can also be defined for scheme A.

EXAMPLE 7.2 (continued). *As we saw, Carbon dioxide and Ventilation can be omitted as explanatory variables in the model. Table 7.9 shows the estimated parameters and their standard errors for this reduced model, while Table 7.10 shows the expected values, the standardized residuals and Cook's distances for each group i.*

LOGISTIC REGRESSION ANALYSIS

TABLE 7.9. ML-estimates for the regression parameters with standard errors for the data in Table 7.3.

Variable	Parameter	Estimate	Standard error
Intercept	β_0	-11.160	4.081
T: Temperature	β_1	1.042	0.284
M: Moisture	β_2	-0.703	0.168
F: Fresh air	β_4	0.950	0.171
D: Dust	β_5	17.531	6.085

TABLE 7.10. Standardized residuals and Cook's distances for a logistic regression model with only four explanatory variables for the indoor climate data. Also shown are the values of the explanatory variables.

Group	Observed number	Expected number	Explanatory variables				Standardized residuals	Cook's distances
i	x_i	$n_i \pi_i$	T	M	F	D	r_i	C_i
1	18	17.21	22.0	30	8.5	0.20	1.37	1.83
2	16	14.10	21.5	25	6.1	0.08	1.37	0.54
3	4	6.25	21.5	25	4.8	0.06	-1.64	0.83
4	13	11.26	18.5	25	9.2	0.07	1.72	2.31
5	12	11.98	20.0	25	8.7	0.08	0.02	0.00
6	4	5.41	20.0	25	5.2	0.12	-1.18	0.59
7	14	16.29	20.5	30	13.1	0.09	-3.12	0.60
8	18	17.64	21.0	30	12.5	0.06	0.39	0.02
9	9	6.62	21.5	30	8.7	0.07	1.71	0.73
10	8	9.23	21.0	30	9.3	0.09	-0.78	0.12

Figure 7.3 shows the standardized residuals in Table 7.10 plotted against the four explanatory variables included in the final model. Since the model describes the data, we should expect few residuals outside the range (-2,+2) and a random scattering of the residuals within these limits.

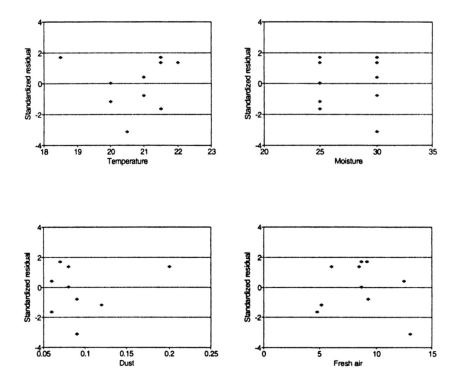

FIGURE 7.3. Residual diagrams for the indoor climate data with the standardized residuals plotted against the four explanatory variables.

The four diagrams in Figure 7.3 show that only one residual is somewhat extreme. It is group 7 with a relatively high expected value as compared with the observed count of students feeling a pleasant indoor climate. For most of the explanatory variables this group does not correspond to an extreme value, except maybe for Fresh air, where group 7 has the largest value, but group 6 has almost as large a value.

The Cook distances in Table 7.10 shows that two groups have Cook distances, which are clearly larger than the other 8 values, namely group 1 and 4. Figure 7.3 shows immediately the reasons for this. The residual diagram for Temperature shows that group 4 has a temperature that is well below the lump of the other temperatures. Group 1 on the other hand stands out in the diagram for Dust with a much higher degree of dust that the other groups. The reason for Cook's distance to be large for groups 1 and 4 is thus that these two groups are connected with extreme values of the explanatory variables which of course, if removed from the data set, will influence the estimates more than if groups with not so extreme values are removed. This example also shows that a Cook distance can be large even when the residual is moderate in value.

If diagnostics point to special cases or groups as having a large influence on the parameter estimates or on the model fit, one often used technique is to remove these groups and re-estimate the parameters and retest the fit of the model. This does not mean that it is always justified to remove observations from a data set. Only if it becomes clear, by careful checking of the data collection procedure, that errors have occurred it can be fully justified to correct these errors and thus change the data. The influence of one of the cases or groups may on the other hand be so invalidating for an otherwise simple and reasonable model for the data, that a sound way to report one's findings is to say, that the simple model fits the data except for some unexplainable extreme observations connected with certain groups. In many cases further investigations may shown that such extreme groups, for reasons not clear at present, are bound to be observed. With all these reservations in mind it can be recommended to repeat the estimations and test without groups pointed out extremes by well chosen diagnostics like standardized residuals and Cook distances.

EXAMPLE 7.2 (continued). *Table 7.11 represents a summary of the changes in the parameter estimates and the Z-test statistics for goodness of fit of the model, when in turn groups 1 and 4, with large Cook distances, and group 7 with a standardized residual numerically larger than 3, are removed from the data set.*

TABLE 7.11. Parameter estimates and observed values (z) of the goodness of fit test statistics Z for a logistic regression model, with degrees of freedom (df) and significance levels (p), when all groups are included, and with groups 1, 4 and 7 removed.

	Parameter estimates			
Variable	All groups	Without group 1	Without group 4	Without group 7
Intercept	-11.16	-6.98	-20.92	-14.20
Temperature	1.04	0.83	1.50	1.28
Moisture	-0.70	-0.62	-0.70	-0.85
Fresh air	0.95	0.84	0.94	1.21
Dust	17.53	2.96	19.79	19.96
z	10.30	8.36	7.33	3.69
df	5	4	4	4
p	0.067	0.079	0.120	0.449

The table shows, as expected, that exclusion of group 1 change the parameter esti-

mates in particular the intercept and the β for Dust. Also the exclusion of group 4 change the parameter estimates, here mostly the β for Temperature and the intercept. The exclusion of group 7 does change the parameter estimates much, but in contrast to the two other groups, where the fit of the model is not significantly improved, the fit is markedly improved if we remove group 7 with the large residual.

7.7 Predictions

For given values $\mathbf{z} = (z_1,...,z_p)$ of the explanatory variables one can - this might indeed be the whole purpose of the data analysis - make a prediction of the probabilities π_v, which are the basis for the expected value of Y_v. Since

$$E[Y_v] = \pi(\mathbf{z}) \cdot 1 + (1 - \pi(\mathbf{z})) \cdot 0 = \pi(\mathbf{z}),$$

when the explanatory variables corresponding to Y_v are $\mathbf{z} = (z_1, ... ,z_p)$, the connection between the prediction probabilities $\pi(\mathbf{z})$ and the z's is

$$\ln\left(\frac{\pi(\mathbf{z})}{1-\pi(\mathbf{z})}\right) = \beta\mathbf{z},$$

From (7.15) we the get

$$\text{var}\left[\ln\left(\frac{\hat{\pi}(\mathbf{z})}{1-\hat{\pi}(\mathbf{z})}\right)\right] = \mathbf{z}'(\mathbf{ZWZ}')^{-1}\mathbf{z}. \tag{7.34}$$

Confidence limits for

$$\text{logit}(\pi(\mathbf{z})) = \ln\left(\frac{\pi(\mathbf{z})}{1-\pi(\mathbf{z})}\right)$$

with level of confidence $1 - \alpha$ are, therefore, given by

$$\text{logit}(\hat{\pi}(\mathbf{z})) \pm u_{1-\alpha/2} \cdot \hat{\sigma},$$

where

$$\hat{\sigma}^2 = \mathbf{z}'(\mathbf{Z\hat{W}Z}')^{-1}\mathbf{z}.$$

The confidence limits for the prediction probabilities $\pi(\mathbf{z})$ are thus given by

$$\frac{\exp\left(\text{logit}(\hat{\pi}(\mathbf{z}))+u_{1-\alpha/2}\cdot\hat{\sigma}\right)}{1+\exp\left(\text{logit}(\hat{\pi}(\mathbf{z})+u_{1-\alpha/2}\cdot\hat{\sigma}\right)}$$

and

$$\frac{\exp\left(\text{logit}(\hat{\pi}(\mathbf{z}))-u_{1-\alpha/2}\cdot\hat{\sigma}\right)}{1+\exp\left(\text{logit}(\hat{\pi}(\mathbf{z})-u_{1-\alpha/2}\cdot\hat{\sigma}\right)}.$$

EXAMPLE 7.2 (continued). *We shall now show the predicted response probabilities and their upper and lower 95% confidence limits for the indoor climate data, when only the four significant explanatory variables are included in the model. Table 7.12 shows the predicted response probabilities for all 10 groups together with the upper and lower 95% confidence limits.*

TABLE 7.12. The predicted response probabilities for all 10 groups together with the upper and lower 95% confidence limits.

Group	\multicolumn{4}{c}{Explanatory variables}	Predicted probability	Upper and lower Confidence limits
	T	M	F	D		
i=1	22.0	30	8.5	0.20	0.906	(0.709 , 0.974)
2	21.5	25	6.1	0.08	0.706	(0.542 , 0.829)
3	21.5	25	4.8	0.06	0.329	(0.194 , 0.499)
4	18.5	25	9.2	0.07	0.626	(0.421 , 0.793)
5	20.0	25	8.7	0.08	0.855	(0.739 , 0.925)
6	20.0	25	5.2	0.12	0.301	(0.162 , 0.488)
7	20.5	30	13.1	0.09	0.958	(0.889 , 0.985)
8	21.0	30	12.5	0.06	0.929	(0.824 , 0.973)
9	21.5	30	8.7	0.07	0.414	(0.259 , 0.588)
10	21.0	30	9.3	0.09	0.513	(0.326 , 0.661)

The number 0.906 in row one of Table 7.12 is computed in the following way: For the values 22.0, 30, 8.5 and 0.20 and the β's in Table 7.9, we get the logit

$$-11.16 + 22.0 \cdot 1.042 + 30 \cdot (-0.703) + 8.5 \cdot 0.950 + 0.20 \cdot 17.531 = 2.2552 .$$

Hence

$$\hat{\pi}_1 = \frac{e^{2.2552}}{1 + e^{2.2552}} = \frac{9.5372}{10.5372} = 0.905 ,$$

which is 0.906 apart from rounding errors.

7.8 Dummy variables

When we formulated the logit-model as a regression model in section 6.3, we in principle only treated binary explanatory variables. If, however, in a logit-model

$$\text{logit}(\pi_{1|jkl}) = 2\tau_1^A + 2\tau_{1j}^{AB} + 2\tau_{1k}^{AC} + 2\tau_{11}^{AD} , \qquad (7.35)$$

one of the explanatory variables, for example variable D, has more than two levels, we can not directly formulate the model (7.35) as

$$\log(\pi_{1|jkl}) = \beta_0 + \beta_1 z_j + \beta_2 z_k + \beta_3 z_l, \qquad (7.36)$$

with $z_l = 1$ or -1. If we want to formulate (7.35) as a regression model, we need as many regression parameters β as there are unconstrained τ_{11}^{AD}'s. Thus let variable D have 5 levels. Then there are 4 unconstrained τ_{11}^{AD}'s and we need 4 β's and, therefore, also 4 constructed explanatory variables $z_l^{(1)}$, $z_l^{(2)}$, $z_l^{(3)}$ and $z_l^{(4)}$. The way to obtain this is known as the **dummy variable method** and the z's are called dummy variables.

One way to construct dummy variables is shown in Table 7.13.

TABLE 7.13. Definition of dummy variables for a polytomous explanatory variable with five levels.

Category	$z_l^{(1)}$	$z_l^{(2)}$	$z_l^{(3)}$	$z_l^{(4)}$
l=1	1	0	0	0
2	0	1	0	0
3	0	0	1	0
4	0	0	0	1
5	-1	-1	-1	-1

If we define $\beta_{31}, \ldots, \beta_{34}$ as

$$\beta_{31} = \tau_{11}^{AD}$$

$$\beta_{32} = \tau_{12}^{AD}$$

$$\beta_{33} = \tau_{13}^{AD}$$

$$\beta_{34} = \tau_{14}^{AD}$$

we can change (7.35) with L=5 to

$$\ln(\pi_{1|jkl}) = \beta_0 + \beta_1 z_j + \beta_2 z_k + \beta_{31} z_l^{(1)} + \beta_{32} z_l^{(2)} + \beta_{33} z_l^{(3)} + \beta_{34} z_l^{(4)}. \qquad (7.37)$$

Models (7.35) and (7.37) are in fact identical. This means that if we use dummy variables, there is thus no reduction in the number of parameters in the regression model, and estimation of the parameters and a goodness of fit test based on (7.35) will tell us if a logistic regression model can describe the data at all.

LOGISTIC REGRESSION ANALYSIS 185

To see that (7.35) and (7.37) are identical, we choose to check the values of the last 4 terms in (7.37) for the typical category l=1 and the special category l=5. For l=1, we get, according to Table 7.13

$$\beta_{31}z_1^{(1)}+\beta_{32}z_1^{(2)}+\beta_{33}z_1^{(3)}+\beta_{34}z_1^{(4)} = \beta_{31} = 2\tau_{11}^{AD},$$

which is the last term in (7.35) for l=1. For l=5, we get

$$\beta_{31}z_5^{(1)}+\beta_{32}z_5^{(2)}+\beta_{33}z_5^{(3)}+\beta_{34}z_5^{(4)}$$

which according to Table 7.13 is equal to

$$-\beta_{31}-\beta_{32}-\beta_{33}-\beta_{34} = -2\tau_{11}^{AD}-2\tau_{12}^{AD}-2\tau_{13}^{AD}-2\tau_{14}^{AD} = 2\tau_{15}^{AD},$$

due to the constraint

$$\sum_{l=1}^{5}\tau_{1l}^{AD} = 0.$$

Thus for l=1 and l=5, (7.35) and (7.37) are identical.

When dummy variables are derived from the logit-model, as is done here, the correct definition of the variables is the one given in Table 7.13. It is, however, much more common (for example in econometrics) to define dummy variables with reference to a **base line category**, often the last response category. The values of the dummy variables for the base line category are all 0, so that the dummy variables for 5 categories are defined by Table 7.14.

TABLE 7.14. Definition of dummy variables for an explanatory variable with five levels, number five being the base line category.

Category	$z_l^{(1)}$	$z_l^{(2)}$	$z_l^{(3)}$	$z_l^{(4)}$
l=1	1	0	0	0
2	0	1	0	0
3	0	0	1	0
4	0	0	0	1
5	0	0	0	0

It is easy to check, that if we change from the dummy variables defined in Table 7.13 to the dummy variables defined in Table 7.14, all the β's are changed by the additive constant

$$\beta_{35} = -\beta_{31}-\beta_{32}-\beta_{33}-\beta_{34}.$$

It is added to β_0 and subtracted from β_{31} through β_{34}.

In some situations the scores are preassigned to the levels of a polytomous explanatory variable. If for variable D being the polytomous variable, $(z_{31}, \ldots, z_{3L})$ are the scores preassigned to the L levels, the logistic regression model becomes

$$\log(\pi_{1|jkl}) = \beta_0 + \beta_1 z_j + \beta_2 z_k + \beta_3 z_{3l}, \qquad (7.38)$$

but we have no guarantee that the scoring of the levels of variable D is optimal in the sense that we get as good a fit as for the model with dummy variables. In the example below, we shall demonstrate this. But before discussing the example, we have to clarify the connection between model (7.37), using dummy variables, and model (7.38), using a fixed scoring of the response categories. The best way to do this is to demonstrate, that if we use a scoring, which is proportional to the maximum likelihood estimates for the dummy variables, then the β_3's estimated from (7.38) are proportional to the β's estimated from (7.37). Such a result can be paraphrased by saying that the β-estimates obtained using dummy variables represent an "optimal scoring" of the categories of explanatory variable D. The argument runs as follows.

Suppose that we in model (7.38) use $z_{3l} = c\hat\beta_{3l}$, $l=1,...,L$, where $\hat\beta_{3l}$ are the ML-estimates obtained from the dummy variable model (7.37), as the scoring of the response categories, then the last term in (7.38) becomes

$$\beta_3 z_{3l} = \beta_3 c \hat\beta_{3l} \qquad (7.39)$$

But if we replace the β's in (7.37) by their ML-estimates, we get for category l=1 $\hat\beta_{31} z_1^{(1)} = \hat\beta_{31}$. A comparison with (7.39) for l=1 then shows that the ML-estimates for the β's obtained from the dummy variable method are proportional with the ML-estimates for the β's obtained by using the dummy variable β's as scores for the categories in the fixed scoring model (7.38).

EXAMPLE 7.1 (continued). *We return to the data in Table 7.1. In this table, we used a fixed scoring of the age categories, namely the interval midpoints. We now compare this scoring with using dummy variables. Since there are five levels of variable Age, we define four dummy variables Age1, Age2, Age3 and Age4. The dummy variable scoring of the five age intervals is in accordance with Table 7.13. The regression parameter estimates are shown in Table 7.15.*

TABLE 7.15. Estimated regression parameters for the data in Table 7.1 with dummy variables according to Table 7.13.

Variable	Parameter	Estimate	Standard error
Intercept	β_0	-1.629	0.062
Sector	β_1	0.165	0.055
Employment	β_2	0.028	0.051
Age1	β_{31}	-0.163	0.096
Age2	β_{32}	0.211	0.088
Age3	β_{33}	0.211	0.100
Age4	β_{34}	0.081	0.108

If we compare with Table 7.2, the ML-estimates for the Intercept, Sector and Employment has not changed much. The Z-test statistic for the model with dummy variables, however, has observed value

$$z(H) = 10.71, \quad df = 13,$$

with level of significance p = 0.635. Hence the model fit is much better than with the "midpoint" scores (25, 35, 45, 55, 65). The β-estimates in Table 7.15 in fact suggest that the optimal scoring of the age intervals in this group are not the interval midpoints. The sum of the β-estimates for the four age variables in Table 7.15 is 0.340. Hence the last age category must be assigned the value -0.340 in order for the β's to sum to zero. If we choose scores proportional to the estimated β-values from the dummy variable method in such a way, that the range is 40 and the median is 45, as for the age interval midpoints, we get the "optimal" scores (33, 60, 60, 51, 20). If we use these scores for the five age categories, we get the estimates in Table 7.16.

TABLE 7.16. Estimated regression parameters for the data in Table 7.1 with the scoring (33, 60, 60, 51, 20) for the age categories.

Variable	Parameter	Estimate	Standard error
Intercept	β_0	-2.249	0.192
Sector	β_1	0.166	0.054
Employment	β_2	0.028	0.050
Age	β_3	0.014	0.004

Again the estimates have not changed much, except of course for β_0, but as expected the goodness of fit for this model is about the same as for the model with dummy variables, namely

$$z(H) = 10.71, \quad df = 16,$$

with level of significance p = 0.827. That the fit is even slightly better, judged by the level of significance, is because we have "cheated" a little by using the data to construct the "optimal" scoring.

7.9 Polytomous response variables

For polytomous response variables, i.e. if the response variable has more than two possible response categories, we can not use the logistic regression model in the form we have used so far. It is at least not immediately clear to what probabilities the logistic transformation shall be applied. Thus let the possible values of the response variable Y_v be t = 1,...,T and assume that we are in scheme B. For each i, there will then be T observed numbers $x_{i1},...,x_{iT}$, where x_{it} is the number of persons in the group with response $\{Y_v = t\}$ for which the values of the explanatory variables are $z_{1i},...,z_{pi}$.

The distribution of the random variables $X_{i1}, ... ,X_{iT}$ corresponding to the observed values $x_{i1},...,x_{iT}$ is a multinomial distribution

$$(X_{i1},...,X_{iT}) \sim M(n_i,\pi_{i1},...,\pi_{iT}) ,$$

where n_i is the number of cases with values $z_{1i},...,z_{pi}$ of the explanatory variables, and

$$\pi_{it} = P(Y_v = t|z_{1i},...,z_{pi}) .$$

The likelihood function then becomes

$$L = \prod_{i=1}^{I} \binom{n_i}{x_{i1}...x_{iT}} \pi_{i1}^{x_{i1}}...\pi_{iT}^{x_{iT}} . \qquad (7.40)$$

It is possible to formulate a number of regression models based on the probability parameters π_{it} of the likelihood function (7.40). Since there are T-1 unrestricted parameters one possibility is to consider the T-1 **partial logistic regression models**

$$\ln\left(\frac{\pi_{it}}{1-\pi_{it}}\right) = \beta_0^{(t)} + \sum_j \beta_j^{(t)} z_{ji} , \; t = 1,...,T-1 , \qquad (7.41)$$

with one set of regression parameters $(\beta_0^{(t)},...,\beta_p^{(t)})$ for each value of t. This is not necessarily the most obvious way to establish T-1 partial regression models. An attractive alternative is thus to compare the π_{it}'s to the same denominator, which results in the model

$$\ln\left(\frac{\pi_{it}}{1-\pi_{iT}}\right) = \beta_0^{(t)} + \sum_j \beta_j^{(t)} z_{ji} , \; t = 1,...,T-1 . \qquad (7.42)$$

For each t this is a logistic regression model in a conditional sense. Assume that the response can only take the values t and T. Then $\pi_{iT} = 1 - \pi_{it}$ and in the conditional distribution given that t is either t or T (7.42) is for each t a **conditional logistic regression model**.

Another idea is to compare neighbouring probabilities π_{it} and π_{it+1}, giving the model

$$\ln\left(\frac{\pi_{it}}{\pi_{it+1}}\right) = \beta_0^{(t)} + \sum_j \beta_j^{(t)} z_{ji} , \; t = 1,...,T-1 . \qquad (7.43)$$

Model (7.43) is also a conditional logistic regression model if we condition on t being either t or t+1.

A final partial model with some attractive properties is the **continuation regression model** given by

LOGISTIC REGRESSION ANALYSIS

$$\ln\left(\frac{\pi_{it}}{\sum_{s>t} \pi_{is}}\right) = \beta_0^{(t)} + \sum_j \beta_j^{(t)} z_{ji} \ , \ t = 1,\ldots,T-1 \ . \tag{7.44}$$

The idea is here to compare π_{it} in a logistic way with 1 minus the cumulative probability

$$P_{it} = P(S \leq t) = \sum_{s \leq t} \pi_{is} \ , \tag{7.45}$$

where in S the response is a random variable with range $(1, \ldots, T)$.

Note that the β-parameters in models (7.41) to (7.44) are not the same parameters in spite of the common notation.

It is a common advantage for all these models, that we can use the statistical methodology developed for the simple dichotomous logistic regression model.

It is on the other hand a common disadvantage of the models, that each gives rise to not one, but T-1 sets of regression parameters. This means that an explanatory variable may give a significant contribution to the variation of the response variable for one value of t, but not for other values of t. We can thus easily come across situations, as in Example 7.4 below, where a statistical analysis based on one of the models for one value of t clearly indicates that the j'th explanatory variable can be omitted, while the analysis for another value of t equally clearly indicates that the j'th explanatory variable contributes significantly to explain the variation in the response variable.

Models have also been developed, where only one set of regression parameters is enough to describe the contribution of the explanatory variables. The McCullagh-model (cf. the biographical notes) has the form

$$\ln\left(\sum_{s=1}^{t} \pi_{is}\right) - \ln\left(\sum_{s=t+1}^{T} \pi_{is}\right) = \alpha_t + \sum_j \beta_j z_{ji} \ , \ t = 1,\ldots,T-1 \ , \tag{7.46}$$

This model can be described as a logistic regression model for the cumulative probabilities (7.45) since (7.46) is equivalent with

$$\ln\left(\frac{P_{it}}{1-P_{it}}\right) = \alpha_t + \sum_j \beta_j z_{ji} \ .$$

Model (7.46) implies that the contributions of the explanatory variables via

$$\sum_j \beta_j z_{ji} \ ,$$

are distributed over the T categories of the response variables in such a way that the logits of the cumulative probabilities (7.46) increase or decrease with

$\alpha_1,...,\alpha_{T-1}$. It follows that the model contains p parameters $(\beta_1,...,\beta_p)$, which describe the contributions of the explanatory variables, and T-1 parameters $\alpha_1,...,\alpha_{T-1}$, which describe the distribution over the response categories.

A model, which also have only one set of regression parameters is

$$\ln\left(\frac{\pi_{it}}{\pi_{iT}}\right) = \left(\beta_0 + \sum_j \beta_j z_{ji}\right) w_t, \quad t = 1,...,T-1. \qquad (7.47)$$

For this model

$$\pi_{it} = \frac{\exp\left(\beta_0 + \sum_j \beta_j z_{ji}\right) w_t}{\sum_{s=1}^{T} \exp\left(\beta_0 + \sum_j \beta_j z_{ji}\right) w_s}. \qquad (7.48)$$

The probabilities π_{it}, given by (7.48), are invariant under linear transformations of the w's. In fact if $w_t^* = a + bw_t$ then (7.48) becomes

$$\pi_{it}^* = \frac{\exp\left[\left(\beta_0 + \sum_j \beta_j z_{ji}\right)(a+bw_t)\right]}{\sum_{s=1}^{T} \exp\left[\left(\beta_0 + \sum_j \beta_j z_{ji}\right)(a+bw_s)\right]}$$

$$= \frac{\exp\left[a\left(\beta_0 + \sum_j \beta_j z_{ji}\right)\right] \cdot \exp\left[\left(b\beta_0 + \sum_j b\beta_j z_{ji}\right) w_t\right]}{\exp\left[a\left(\beta_0 + \sum_j \beta_j z_{ji}\right)\right] \cdot \sum_{s=1}^{T} \exp\left[\left(b\beta_0 + \sum_j b\beta_j z_{ji}\right) w_s\right]}.$$

which is identical to (7.48) apart from a rescaling of the β's. Hence we can normalize the w's by introducing the constraints

$$\sum_{t=1}^{T} w_t = 0 \quad \text{and} \quad \sum_{t=1}^{T} w_t^2 = 1. \qquad (7.49)$$

From (7.48) then follows by simple algebra that

$$\sum_{t=1}^{T} w_t \cdot \ln\left(\frac{\pi_{it}}{\pi_{iT}}\right) = \beta_0 + \sum_j \beta_j z_{ji}.$$

Model (7.47) can thus be described as a conditional logistic regression model (the conditioning being on either t or T being the response) where a weighted sum of the logarithms of the response probabilities is a regression model.

Both models (7.46) and (7.47) can be checked in the usual way by means of a Z-test statistic for goodness of fit. Z has in this situation the form

$$Z = 2\sum_{i=1}^{I}\sum_{t=1}^{T} X_{it}\left(\ln(X_{it}) - \ln(n_i \hat{\pi}_{it})\right). \quad (7.50)$$

For model (7.46), $\hat{\pi}_{it}$ is given by

$$\hat{\pi}_{it} = \hat{P}_{it} - \hat{P}_{i,t-1},$$

where

$$\hat{P}_{it} = \frac{\exp\left(\hat{\alpha}_t + \sum_j \beta_j z_{ji}\right)}{1 + \exp\left(\hat{\alpha}_t + \sum_j \beta_j z_{ji}\right)},$$

For model (7.47) $\hat{\pi}_{it}$ is given by

$$\hat{\pi}_{it} = \frac{\exp\left[\left(\beta_0 + \sum_j \beta_j z_{ji}\right)w_t\right]}{\sum_{s=1}^{T}\exp\left[\left(\beta_0 + \sum_j \beta_j z_{ji}\right)w_s\right]}.$$

Under models (7.46) and (7.50) is approximately χ^2-distributed with

$$df = I(T-1) - p - (T-1) = (I-1)(T-1) - p$$

degrees of freedom, while (7.50) under model (7.47) is approximately χ^2-distributed with

$$df = I(T-1) - p - 1$$

degrees of freedom.

EXAMPLE 7.4. *Stress at work.*
We return to the question of feeling stressed at work. In Example 7.1 the responses were dichotomized to the categories Feeling stressed and Not feeling stressed. Actually the question: "Do you feel stress at work?" in the Welfare Study had three response categories: "Yes", "Yes, sometimes" and "No". Table 7.17 shows the responses distributed on all three response categories, now with the explanatory variables:

Employment, with categories White collar worker =1 and Blue collar worker =2.
Sector, with categories Private =1 and Public =2.
Geographical region, with categories Copenhagen =1, Zeeland and Fuen =2 and Jutland =3.

TABLE 7.17. The sample in the Danish Welfare Study cross-classified according to Stress at work, Employment, Sector and Geographical region.

\multicolumn{3}{c}{Observed responses on the question: Do you feel stressed at work?}			\multicolumn{3}{c}{Explanatory variables}		
Yes	Yes sometimes	No	Employment	Sector	Region
54	107	97	1	1	1
43	54	68	2	1	1
35	91	113	1	2	1
10	20	45	2	2	1
45	94	104	1	1	2
62	110	178	2	1	2
29	88	103	1	2	2
13	15	39	2	2	2
68	135	125	1	1	3
100	165	291	2	1	3
52	98	139	1	2	3
18	32	87	2	2	3

Source: The data base of the Danish Welfare Study. Cf. Example 3.2.

If we analyze the data in Table 7.17 by the four partial logistic or conditional logistic regression models (7.41) to (7.44), we get the parameter estimates shown in 7.18. Also shown are the observed values of the Z-test statistic and the corresponding levels of significance. Parameters, which are significantly different from 0 at a 95%-level are highlighted by bold types.

TABLE 7.18. Parameter estimates for models (7.41) to (7.44) with observed values (z) of the Z-test statistic, degrees of freedom (df) and levels of significance (p) for the data i Table 7.17.

Model		$\beta_0^{(t)}$	$\beta_1^{(t)}$	$\beta_2^{(t)}$	$\beta_3^{(t)}$	z	df	p
(7.41)	t=1	**-0.928**	-0.055	**-0.320**	-0.035	9.77	8	0.281
	t=2	0.439	**-0.469**	-0.211	-0.006	3.53	8	0.897
(7.42)	t=1	**-1.222**	0.170	-0.083	0.001	3.33	8	0.912
	t=2	-0.340	-0.092	0.040	-0.003	1.41	8	0.994
(7.43)	t=1	**-0.881**	**0.261**	-0.124	0.012	6.78	8	0.557
	t=2	**1.100**	**-0.550**	**-0.340**	-0.072	4.06	8	0.852
(7.44)	t=1	**-0.928**	-0.055	**-0.320**	-0.035	9.77	8	0.281
	t=2	**1.100**	**-0.550**	**-0.340**	-0.072	4.06	8	0.852

It is obvious from this table that we get very different results when applying the different models. It is also clear that the models do not give a unique universal answer to the obvious question: Which explanatory variables contribute significantly to the description of the variation in the response variable? It is indeed the exception rather than the rule that the parameter estimates for t=1 and t=2 are identical, even when both are significantly different from 0. The only clear conclusion is that the explanatory variable Geographical region does not contribute significantly. Stressed work is thus not, as one maybe would have believed, more common in Copenhagen, the only real big city in Denmark, than in the rest of the country, according to peoples own conception.

In order to apply model (7.47) to the data in Table 7.17, we have chosen an equidistant scoring, which with the constraints (7.49) becomes

$$(w_1, w_2, w_3) = (0.71, 0, -0.71).$$

Table 7.19 shows the parameter estimates, the observed values of the Z-test statistics and the corresponding levels of significance for models (7.46) and (7.47). Parameters, which are significantly different from 0 at a 95%-level are, as in Table 7.18, highlighted by bold types.

TABLE 7.19. Parameter estimates for models (7.46) and (7.47) with observed values (z) of the Z-test statistic, degrees of freedom (df) and levels of significance (p) for the data i Table 7.13.

Model	β_0	β_1	β_2	β_3	z	df	p
(7.46)	-0.333	**0.286**	**0.331**	0.054	56.09	20	0.000
(7.47)	-	**0.347**	**0.360**	0.061	32.92	19	0.025

The two models (7.46) and (7.47), which provide one set of β-estimates, both desc-

ribe the data much less satisfactorily than the partial models. The regression parameters β_1, β_2 and β_3 are, on the other hand, almost identical for the two models and as for the partial models, they show that Geographical region does not make a significant contribution.

7.10 Bibliographical notes

Logistic regression analysis was suggested by Cox (1970). It has been known for a long time that the canonical parameter for the binomial distribution is obtained by a logistic transformation of the probability parameter, cf. for example Lehmann (1959) or Barndorff-Nielsen (1978). The conditions for existence of unique solutions to the likelihood equations was given by Albert and Andersson (1984).

The logistic regression model is a special case of the generalized linear model as introduced by Nelder and Wedderburn (1972) and fully discussed in McCullagh and Nelder (1983). It is shown in the book by McCullagh and Nelder that the Newton-Raphson solution of the likelihood equations for generalized linear models - and hence for logistic regression - can be obtained by a method similar to the weighted least squares solution for ordinary regression models. The method is called the **iterative weighted least squares method** or iterative reweighed least square method, cf. also Andersen (1990) section 3.7 or Agresti (1996) section 4.7.3.

Standardized residuals and Cook distances for logistic regression were discussed by Pregibon (1981).

Continuation regression models was suggested by Thompson (1977) and Fienberg and Mason (1979). The McCullagh model was suggested by McCullagh (1980).

Logistic regression analysis is treated in most books on categorical data and analysis and in many books on applied regression analysis. Cf. for example Andersen (1990) chapter 9, Agresti (1996) chapter 4 and Weisberg (1985). Recently special books on logistic regression analysis have been published, for example Hosmer and Lebeshowes (1989).

7.11 Exercises

[For some of the exercises you need a computer package like BMDP or SAS to obtain test statistics, estimates, standard errors, predictions or diagnostics. For readers without access to such packages, selected and organised output from the SAS package is shown in the Appendix]

7.1 The table below shows for 8 hypothetical groups the number of positive responses x_i out of n_i observations together with three sets of two explanatory variables.

Group	Positive response	Observed number	Set 1		Set 2		Set 3	
i	x_i	n_i	z_1	z_2	z_1	z_2	z_1	z_2
1	13	25	0.5	0.5	0.5	3.0	0.5	3.0
2	10	10	3.0	0.5	2.0	2.5	2.0	2.5
3	11	15	1.5	1.5	1.0	1.5	1.0	1.5
4	8	10	2.5	2.5	3.5	1.5	1.5	1.5
5	0	9	2.0	2.5	1.5	1.5	2.5	2.5
6	0	15	1.0	1.5	2.5	2.5	3.5	1.5
7	3	17	3.0	3.0	3.0	0.5	3.0	0.5
8	15	15	3.5	1.5	0.5	0.5	0.5	0.5
9	0	12	0.5	3.0	3.0	3.0	3.0	3.0

For which of the three sets of explanatory variables are there solutions to the likelihood equations?

7.2 In this exercise we consider 10 hypothetical cases with 0 - 1 responses and two explanatory variables. As in exercise 7.1 there are three alternative sets of explanatory variables.
The table below shows for each case the response together with the three different sets of explanatory variables.

Case	Response	Set 1		Set 2		Set 3	
i	y_i	z_1	z_2	z_1	z_2	z_1	z_2
1	0	0.5	3.0	0.5	3.0	2.5	0.0
2	1	2.5	0.0	2.0	2.5	0.5	3.0
3	1	2.5	2.5	1.0	1.5	2.0	2.5
4	1	1.5	1.5	1.5	1.5	1.0	1.5
5	0	2.0	2.5	2.5	2.5	3.5	1.5
6	0	1.0	1.5	3.5	1.5	3.0	0.5
7	1	3.5	1.5	3.0	0.5	1.5	1.5
8	1	3.0	0.5	0.5	0.5	2.5	2.5
9	1	0.5	0.5	3.0	3.0	0.5	0.5
10	1	3.0	3.0	0.0	1.0	0.0	1.0
11	0	0.0	1.0	2.5	0.0	3.0	3.0

7.3 Write the data set in Example 6.2, Table 6.2 (with Party membership as the response variable) as a logistic regression data set, i.e in the same form as Table 7.1.

(a) Estimate the regression parameters and their standard errors.

(b) Carry out a logistic regression analysis to determine which of the three explanatory variables contribute significantly to the description of the variation in the response variable.

(c) Compare the results in (b) with the conclusions drawn in Example 6.1

7.4 The Danish Institute for Building Research in 1983 made an investigation of the indoor climate at a number of City Halls of independent municipalities in the suburbs of Copenhagen. The response variable was - as in Exercise 4.3 - Irritation of the throat. The explanatory variables were A: Dust, measured in a typical office of the building and B: The amount of Ventilation, also measured in a typical office. The table below shows the number of City Hall employees with irritation of the throat together with the total number interviewed and the values of the explanatory variables.

City Hall no.	Number with irritation of the throat	Number interviewed	Dust (mg/m^3)	Ventilation (liter/sec.)
120	105	301	0.27	7.50
41	22	36	0.16	9.20
70	206	317	0.12	8.60
121	137	196	0.11	7.70

(a) Test if a logistic regression model fits the data.

(b) Estimate the regression parameters and test whether it is necessary to include both, just one, or none of the explanatory variables in the model.

In the table below the standardized residuals and the Cook distances are shown for a logistic regression model with both explanatory variables included.

City Hall no.	Dust (mg/m^3)	Ventilation (liter/sec.)	Standardized residuals	Cook distances
120	0.27	7.50	-0.698	68.85
41	0.16	9.20	+0.698	0.13
70	0.12	8.60	-0.698	0.79
121	0.11	7.70	+0.698	1.87

(c) Comment on the information provided by both the values of the standardized residuals and of the Cook distances.

7.5 The table below is typical for the way results from a logistic regression analysis are presented in many publications. The table is from a report published by the Danish Institute for Border Research (the border being the German-Danish border). The response variable is whether the interviewed person has changed jobs within the last ten years with response values 1 for Yes and 0 for No.

Variable	Estimated regression coefficient (standard error)		
	Model I	Model II	Model III
Intercept	1.2831 (0.9010)	4.6945 (0.5783)	5.4697 (0.9560)
Sex: Men(0) Women(1)	-0.2995 (0.2874)	-0.7735 (0.2622)	-0.7735 (0.2622)
Age: Years	-0.0262 (0.0136)	-0.1007 (0.0127)	-0.0989 (0.0165)
Household income (1000 Dkr.)	0.0002 (0.0021)	-0.0036 (0.0024)	
Number of children at home under 18	-0.0157 (0.1256)		
Number of hours per week your wife or husband goes to work	0.0037 (0.0125)		
Length of education: Month more than 10 years	-0.0043 (0.0070)	0.0133 (0.0066)	

(a) Explain the meaning of the entries in the table.

(b) What would your conclusions be as regards the explanatory variables contribution to explain the differences between those who have changed jobs and those who have not.

(c) How would you describe the differences between the three models shown.

7.6 The data set for this exercise is from an comparative investigation of the sports careers, the social careers, the educational careers and the occupational careers of 290 Danish soccer players, who had played professional soccer outside Denmark from 1945 to 1990. A sample of 131 of these players answered a questionnaire. We consider the answers on seven of the questions in the questionnaire as basis for a logistic regression analysis. The response variable has the value 1 if the player answered Yes to the question "Do you feel your career as a professional soccer player has been an economic success?" and 0 if the answer was No. The explanatory variables were:

A: Married, with value 1 if the player was married or had established a permanent relationship before starting his soccer career abroad, and 0 if he was unattached.
B: Money with value 1, if a motive for seeking a soccer career abroad was to earn money, and 0 if money was not mentioned as a motive.
C: Team, with value 1 if the player claimed that he was on his club's best team almost all the time he played abroad, and 0 otherwise.
D: The number of Years he has played professional soccer outside Denmark.
E: The number of different Clubs he has played for while abroad.
F: The number of Games played on the Danish National soccer team before heleft Denmark the first time to play professionally abroad.

The table below shows the values of -2lnL for a number of logistic regression models with different sets of explanatory variables included.

Included variables	-2lnL
A B C D E F	55.137
B C D E F	55.138
A B C D E	55.407
A B C D F	56.452
A B C E F	61.234
A C D E F	61.462
A B D E F	62.527
B C D E	55.413
B C D F	56.486
B C E F	61.248
C D E F	61.509
B D E F	63.112
A B C D	56.506
A B C E	61.575
A C D E	62.238
A B D E	63.447
A B C F	61.299
A C D F	62.204
A B D F	63.692
A C E F	68.951
A B E F	73.975
A D E F	68.460
B C D	56.550
B C	61.750
C	71.791

(a) Use this table to select those explanatory variables which contribute significantly to describe whether a player feels he has had economic success or not.

(b) Make a table of test statistics, their degrees of freedom and their levels of significance based on which your selection in (a) can be justified.

The table below shows the estimated parameters and their standard errors for the model with explanatory variables B, C and D included.

Variable	Estimate	Stand. error
Intercept	-1.51	0.69
B: Money	1.66	0.71
C: Team	2.12	0.78
D: Years	0.35	0.18

(c) Estimate the probability of economic success if a player has been on the best Team almost all the time, a motive for leaving Denmark was to earn money and he has played 5 years abroad. Estimate also the probability if he was on the best team almost all the time, but he did not give earning money as a motive and he played only 1 year abroad before he returned to Denmark.

7.7 In 1982 the Danish Institute for Border Research investigated the labour mobility in the border region between Denmark and Germany. The persons interviewed were asked if they had been unemployed within the last 5 years. The answers are shown together with the explanatory variables A: Sex and B: Age in the table below.

Sex	Age	Unemployed last 5 years		Sex	Age	Unemployed last 5 years	
		Yes	No			Yes	No
	18-19	1	4		18-19	0	2
	20-24	15	14		20-24	12	10
	25-29	12	19		25-29	8	15
	30-34	10	25		30-34	3	16
Men	35-39	4	28	Women	35-39	5	23
	40-44	6	18		40-44	1	14
	45-49	6	15		45-49	2	12
	50-54	2	15		50-54	3	18
	55-59	3	19		55-59	1	6
	60-	1	13		60-	1	1

(a) Test if a logistic regression model fit these data with Sex scored 1 for Men and 0 for Women, and Age scored in the interval midpoints: 19, 22.5, 27.5, 32.5, 37.5, 42.5, 47.5, 52.5, 57.5 and 65.

(b) Estimate the parameters and determine if both, just one, or none of the explanatory variables are needed to explain the differences between persons with an unemployment history and persons without.

(c) Is it possible for this data set to use the dummy variable method to avoid arbitrarily assigning the age midpoints as scores? If yes, test the new model and estimate its parameters.

7.8 From the data base of the Danish Welfare Study we consider the response variable: Sports activity with response categories 1: At least once a week and 0: Never or occasionally. (Sports activities include swimming, cycling or running to stay fit.) The explanatory variables were:

    A: Sex, with categories Male and Female.
    B: Age with five age intervals 20-29, 30-39, 40-49, 50-59 and 60-69.
    C: Employment sector with categories: Private or Public.
    D: Urbanization with categories Copenhagen, Suburbs of Copenhagen, the Three largest cities (outside Copenhagen), Other cities and Countryside.

The scoring of the two polytomous explanatory variables are: For Age the interval midpoints 25, 35, 45, 55 and 65. For Urbanization Copenhagen = 1, Suburbs = 2, Three largest cities = 3, Other cities = 4 and Countryside = 5.

With these scoring the test statistics $z(H_{(j)})$ defined by (7.30), here just called z, are shown for a selection of included explanatory variables in the table below. The test statistic in line one with all explanatory variables included is the one defined by (7.22).

Explanatory variables included	Explanatory variables excluded	z
A B C D	None	121.16
B C D	A	1.29
A C D	B	111.30
A B D	C	16.40
A B C	D	1.55
C D	A B	111.55
B D	A C	16.21
B C	A D	2.68
C	A B D	111.94
B	A C D	18.57
None	A B C D	123.50

There are 100 possible combinations of categories for the explanatory variables, but for 8 of these either x or n-x are zero, so the number of binomial distributions involved is I = 92.

(a) Determine the number of degrees of freedom for each of the z-values in the table above.

(b) Use the table to determine which explanatory variables contribute significantly to describe the variation in the response variable.

As an alternative we can introduce the four dummy age variables B1, B2, B3 and B4 with definition given in Table 7.13. The table of test statistics now becomes.

Explanatory variables included	Explanatory variables excluded	z
A B1 B2 B3 B4 C D	None	106.22
B1 B2 B3 B4 C D	A	0.81
A B1 B2 B3 B4 D	C	14.53
A B1 B2 B3 B4 C	D	1.47
A C D	B1 B2 B3 B4	125.43
B1 B2 B3 B4 D	A C	15.35
B1 B2 B3 B4 C	A D	2.18
C D	A B1 B2 B3 B4	126.49
B1 B2 B3 B4	A C D	16.82
C	A B1 B2 B3 B4 D	126.87

(c) Determine the degrees of freedom for all z-values in this table and test which explanatory variables contribute significantly to explain the variation in the response variable.

The table below shows the regression parameters estimated both with and without dummy variables together with their standard errors.

	Model with age scored			Model with dummy age variables		
Expl. var.	β	Std. error	Expl. var.	β	Std. error
Intercept	-0.368	0.248	Intercept	1.387	0.208
Sex	0.098	0.087	Sex	0.078	0.087
Age	0.037	0.004	B1	-0.628	0.081
			B2	-0.606	0.080
			B3	0.047	0.097
			B4	0.523	0.112
Sector	-0.360	0.089	Sector	-0.343	0.090
Urbanization	0.040	0.032	Urbanization	0.039	0.032

(d) Compare the estimates in this table and draw your conclusions.

Chapter 8

Association Models

8.1 Introduction

In this chapter we shall discuss a number of association models for two-way tables. If the independence model

$$\tau_{ij} = \tau_{i.}\tau_{.j}$$

fails to fit the data, or for the log-linear parametrization, the hypothesis

$$H: \tau_{ij}^{AB} = 0,$$

is rejected, we often want to search for a model which can describe the dependencies. Such models are called **association models**.

It is beyond the scope of this book to discuss the various association models in very much detail. For a more complete discussion, the reader is referred to Andersen (1990), chapters 10 and 11. We shall, however, give a short review of the most important association models.

8.2 Symmetry models

A very simple association model is the **symmetry model**, according to which the cell probabilities π_{ij} satisfy the relationships

$$H_S: \mu_{ij} = \mu_{ji}. \tag{8.1}$$

The model is thus that the expected values $n\pi_{ij}$ are equal for cells symmetric in relation to the diagonal. The model does not put restrictions on the expected values in the diagonal. The symmetry hypothesis is of course only meaningful for **square contingency tables**, i.e. two-way tables with $I = J$.

The symmetry model is log-linear, since the log-likelihood function can be written

$$\ln L = \text{const.} + \sum_i \sum_j x_{ij} \ln \pi_{ij}$$
$$= \text{const.} + \sum_{i<j} (x_{ij} + x_{ji}) \ln \pi_{ij} + \sum_i x_{ii} \ln \pi_{ii} \ . \tag{8.2}$$

From (8.2) we can conclude that the likelihood equations are

$$x_{ij} + x_{ji} = E[X_{ij}] + E[X_{ji}] = 2n\pi_{ij} = 2\mu_{ij} \ , \ i \neq j$$

and

$$x_{ii} = E[X_{ii}] = n\pi_{ii} = \mu_{ii} \ ;$$

so the common estimate for μ_{ij} outside the diagonal is

$$\hat{\mu}_{ij} = \frac{x_{ij} + x_{ji}}{2} \tag{8.3}$$

and for the diagonal we get

$$\hat{\mu}_{ii} = x_{ii} \ . \tag{8.4}$$

Note here that we estimate the μ-parameters rather than the π's.

The Z-test statistic for the symmetry hypothesis has the form

$$Z(H_S) = 2 \sum_i \sum_j X_{ij} \left(\ln X_{ij} - \ln \hat{\mu}_{ij} \right) \ ,$$

where $\hat{\mu}_{ij}$ is given by (8.3) and (8.4). It follows from $Z(H_S)$ that there are no contributions to the Z-test statistic from the observed counts in the diagonal. Since the model is log-linear $Z(H_S)$ is approximately χ^2-distributed. The number of degrees of freedom is

$$df(H_S) = \frac{I(I-1)}{2} \ ,$$

since the equalities (8.1) defining the model impose a number of restrictions on the log-linear parameters, which is equal to the number of cells over (or below) the diagonal in a square table of dimension I×I. We accordingly reject the symmetry model as a description of a square contingency table if the observed value $z(H_S)$ of $Z(H_S)$ is larger than a suitable percentile, for example the 95% percentile, in a χ^2-distribution with $I(I-1)/2$ degrees of freedom.

Since the mean value in cell (ij) under a log-linear model has the form

$$\mu_{ij} = \exp\left(\tau_{ij}^{AB} + \tau_i^A + \tau_j^B + \tau_0\right)$$

it is clear that the symmetry model (8.1) requires that τ_{ij}^{AB} be symmetric and also that $\tau_i^A = \tau_i^B$. If we relax the last condition and only require τ_{ij}^{AB} to be symmetric, we get the hypothesis of **quasi-symmetry**.

The quasi-symmetry model is thus a log-linear model for a square contingency table under the hypothesis

$$H_{QS} : \tau_{ij}^{AB} = \tau_{ji}^{AB} . \tag{8.5}$$

Also this model is log-linear and the ML-estimates for the log-linear parameters are easy to derive. The Z-test statistic for quasi-symmetry is

$$Z(H_{QS}) = 2\sum_i \sum_j X_{ij}\left(\ln X_{ij} - \ln \hat{\mu}_{ij}\right) , \tag{8.6}$$

where the $\hat{\mu}_{ij}$'s are now the expected numbers under the quasi-symmetry hypothesis (8.5). $Z(H_{QS})$ is approximately χ^2-distributed. In order to determine the number of degrees of freedom, note that the unconstrained 2-factor interactions form a square table of dimension (I-1)×(I-1). Under H_{QS} elements over the diagonal are equal to the elements under the diagonal in this table. Hence the number of degrees of freedom for the Z-test statistic $Z(H_{QS})$ is equal to the number of elements over (or below) the diagonal in a (I-1)×(I-1) table, or

$$df(H_{QS}) = \frac{(I-1)(I-2)}{2} .$$

The hypothesis of quasi-symmetry is rejected at level α if the observed value $z(H_{QS})$ of (8.6) is larger than the (1-α)-percentile $\chi^2_{1-\alpha}(df(H_{QS}))$ in the approximating χ^2-distribution.

Under the hypothesis of quasi symmetry, the likelihood function becomes

$$\ln L = \sum_{i<j}(x_{ij}+x_{ji})\tau_{ij}^{AB} + \sum_i x_{ii}\tau_{ii}^{AB} + \sum_i x_{i.}\tau_i^A + \sum_j x_{.j}\tau_j^B + x_{..}\tau_0 . \tag{8.7}$$

The likelihood equations are therefore

$$x_{ij} + x_{ji} = E[X_{ij}] + E[X_{ji}] = n\pi_{ij} + n\pi_{ji} , \quad i = 1,\ldots,I-1 , \; j = i+1,\ldots,J ,$$

$$x_{ii} = E[X_{ii}] = n\pi_{ii} , \quad i = 1,\ldots,I ,$$

$$x_i. = E[X_i.] = n\pi_i. \quad , i = 1,...,I$$

and

$$x_{\cdot j} = E[X_{\cdot j}] = n\pi_{\cdot j} \quad , j = 1,...,J .$$

These likelihood equations can be solved by a method similar to the iterative proportional fitting procedure, which will produce ML-estimates for the expected cell counts $n\pi_{ij}$. This is all we need to test the goodness of fit of the model by (8.6). If we also need estimates of the τ's, they can be obtained from the log-linear parametrization

$$\ln(\mu_{ij}) = \tau_{ij}^{AB} + \tau_i^A + \tau_j^B + \tau_0 .$$

by inserting the ML-estimates on the left hand side and solving the equations.

EXAMPLE 8.1. *Social group over generations.*
In a follow up to the Danish Welfare Study the social group of the married women in the sample between the age of 40 and 59 was compared to the social group of their fathers. The resulting cross-classification is shown as Table 8.1.

TABLE 8.1 The Social groups of the married women between the age of 40 and 59 in the Danish Welfare Study cross-classified with the Social group of their fathers.

Daughter's social group

Father's social group	I-II	III	IV	V	Total
I-II	12	17	22	3	54
III	8	33	85	95	221
IV	11	26	72	87	196
V	2	18	50	111	181
Total	33	94	229	296	652

Source: The data base from the Danish Welfare Study. Cf. Example 3.2.

The expected numbers under a model of complete symmetry are shown in Table 8.2.

TABLE 8.2. Expected numbers under the model of complete symmetry for the data in Table 8.1.

	Daughter's social group				
Father's social group	I-II	III	IV	V	Total
I-II	12.00	12.50	16.50	2.50	43.50
III	12.50	33.00	55.50	56.50	157.50
IV	16.50	55.50	72.00	68.50	212.50
V	2.50	56.50	68.50	111.00	238.50
Total	43.50	157.50	212.50	238.50	652.00

The observed value of the Z-test statistic is

$$z = 107.96,$$

which with df = 6 degrees of freedom has a level of significance less than 0.0005, so a model of complete symmetry does not fit the data.

The expected numbers under a model of quasi-symmetry are shown in Table 8.3.

TABLE 8.3 Expected numbers under the model of quasi-symmetry for the data in Table 8.1.

	Daughter's social group				
Father's social group	I-II	III	IV	V	Total
I-II	12.00	12.95	24.84	4.20	54.00
III	12.05	33.00	82.05	93.90	221.00
IV	8.16	28.95	72.00	86.90	196.00
V	0.80	19.10	50.10	111.00	181.00
Total	33.00	94.00	229.00	296.00	652.00

The observed value of the Z-test statistic is now

$$z = 6.07,$$

which with 3 degrees of freedom has level of significance $p = 0.108$. The fit of the model is thus satisfactory. Symmetry in the table of expected values means that a movement of a daughter to a lower social group as compared to her father is as likely as a movement to the corresponding higher social group. That we rejected the hypothesis of complete symmetry and accepted the hypothesis of quasi-symmetry thus means that the movements have been symmetric, except for the fact

ASSOCIATION MODELS

that in the marginal distributions over social groups, the daughters still tend to belong to lower social groups, as the marginals in Table 8.3 shows.

8.3 Marginal homogeneity

For square tables one can also consider the hypothesis that the expected marginals are equal, or

$$H_M : \mu_{i\cdot} = \mu_{\cdot i} \, , \, i = 1,\ldots,I \, . \quad (8.8)$$

The model under the hypothesis H_M of **marginal homogeneity** is not log-linear and more general iterative methods than the iterative proportional fitting method are necessary to solve the likelihood equations. Let $\hat{\mu}_{ij}$ be the ML-estimate for μ_{ij} under the restriction (8.8). The Z-test statistic for testing H_M is then

$$Z(H_M) = 2 \sum_i \sum_j X_{ij} \left(\ln X_{ij} - \ln \hat{\mu}_{ij} \right) . \quad (8.9)$$

The number of degrees of freedom $df(H_M)$ for (8.9) is

$$df(H_M) = I - 1 \, ,$$

since the hypothesis (8.8) imposes I-1 constraints on the multinomial parameters, namely

$$\pi_{i\cdot} = \pi_{\cdot i} \, , \, i = 1,\ldots,I-1 \, .$$

The last equation for $i = I$ is not needed since $\pi_{\cdot\cdot} = 1$.

EXAMPLE 8.1 (continued). *Table 8.4 shows the expected numbers for daughter's and father's social group under the model of marginal homogeneity.*

TABLE 8.4. Expected numbers under the model of marginal homogeneity for the data in Table 8.1.

Father's social group	Daughter's social group				
	I-II	III	IV	V	Total
I-II	12.00	17.14	15.19	1.79	46.12
III	7.93	33.00	58.35	56.30	155.58
IV	19.95	47.87	72.00	70.70	210.51
V	6.23	57.57	64.98	111.00	239.79
Total	46.12	155.58	210.51	239.79	652.00

The observed value of the Z-test statistic is

$$z = 101.27,$$

which with df = 3 degrees of freedom has a level of significance less than 0.0005. The model of marginal homogeneity does not fit the data either. This confirms our finding in section 8.2 that the marginal distribution over social groups of the daughter's is different from the marginal distribution over social groups of their fathers.

If the table is a 2×2 table, the hypotheses of symmetry and marginal homogeneity are identical. The Pearson test statistic Q in this case takes a particularly simple form. The cells (1,1) and (2,2) do contribute, while the expected values in cells (1,2) and (2,1) are both $(X_{12} + X_{21})/2$, but

$$X_{12} - \frac{X_{12}+X_{21}}{2} = \frac{X_{12}-X_{21}}{2} = -\left(X_{21} - \frac{X_{12}+X_{21}}{2}\right).$$

Hence Pearson's Q becomes

$$2 \frac{\left(\frac{X_{12}-X_{21}}{2}\right)^2}{\frac{X_{12}+X_{21}}{2}} = \frac{(X_{12}-X_{21})^2}{X_{12}+X_{21}}.$$

This special test statistic is called **McNemars test statistic**.

8.4 RC-association models

As we mentioned in section 8.1, the hypothesis of independence in a two-way contingency table can be formulated in terms of the 2-factor interactions as

$$H : \tau_{ij}^{AB} = 0 .$$

Hence if we want to describe dependencies, the parameters to model are the τ_{ij}'s. Around 1980 Goodman (see biographical notes) introduced the following model for the 2-factor interactions

$$\tau_{ij}^{AB} = \rho \varepsilon_i \delta_j \qquad (8.10)$$

under the name **RC-association model**. Here "RC" stands for Row-Column. The full RC-association model is thus

ASSOCIATION MODELS

$$\ln \mu_{ij} = \rho \varepsilon_i \delta_j + \tau_i^A + \tau_j^B + \tau_0 \ . \tag{8.11}$$

Since the left hand side in (8.10) sums to zero over both indices, we must also have

$$\sum_i \varepsilon_i = \sum_j \delta_j = 0 \ . \tag{8.12}$$

In addition the term $\rho \varepsilon_i \delta_j$ is undetermined up to multiplicative factors. These are resolved by introducing the constraints

$$\sum_i \varepsilon_i^2 = \sum_j \delta_j^2 = 1 \ . \tag{8.13}$$

Finally we have the "old" constraints

$$\sum_i \tau_i^A = \sum_j \tau_j^B = 0 \ . \tag{8.14}$$

Equation (8.11) together with the constraints (8.12), (8.13) and (8.14) define the RC-association model.

Since the model is not log-linear, the likelihood equations are not as simple as for other models, we have met. Since, however, the log-likelihood function can be written as

$$\ln L = \text{const.} + \sum_i \sum_j x_{ij} \rho \varepsilon_i \delta_j + \sum_i x_{i.} \tau_i^A + \sum_j x_{.j} \tau_j^B + x_{..} \tau_0 \ , \tag{8.15}$$

we can derive the likelihood equations by taking partial derivatives of (8.15) with respect to all the parameters. In this way we get the "old" equations

$$x_{i.} = E[X_{i.}] = \mu_{i.} \tag{8.16}$$

and

$$x_{.j} = E[X_{.j}] = \mu_{.j} \tag{8.17}$$

We then get new likelihood equations derived from differentiation with respect to ρ, ε_i and δ_j. If we differentiate partially with respect to ε_i, we get

$$\sum_j \delta_j x_{ij} = \sum_j \delta_j \mu_{ij} \ , \ i=1,\ldots,I \ , \tag{8.18}$$

and if we differentiate partially with respect to δ_j, we get

$$\sum_i \varepsilon_i x_{ij} = \sum_i \varepsilon_i \mu_{ij} \quad , j=1,\ldots,J \ . \tag{8.19}$$

From the partial derivative with respect to ρ we get

$$\sum_i \sum_j \varepsilon_i \delta_j x_{ij} = \sum_i \sum_j \varepsilon_i \delta_j \mu_{ij} \ ,$$

but this equation is satisfied if either (8.18) or (8.19) are satisfied.

Although Equations (8.16) to (8.19) can not be solved by the iterative proportional fitting method, an algorithm (see bibliographical notes) was suggested by Goodman, which is very similar to the algorithm used in the iterative proportional fitting method. If the data is not too ill-behaved it converges quickly to final values.

If $\hat{\mu}_{ij}$ are the estimated expected numbers, i.e.

$$\hat{\mu}_{ij} = \exp\left(\hat{\rho}\hat{\varepsilon}_i\hat{\delta}_j + \hat{\tau}_i^A + \hat{\tau}_j^B + \hat{\tau}_0\right) \ ,$$

where a "^" denotes a ML-estimate, the hypothesis H_{RC} that the data fits an RC-association model can be tested by the Z-test statistic

$$Z(H_{RC}) = 2\sum_i \sum_j X_{ij}\left(\ln X_{ij} - \ln \hat{\mu}_{ij}\right) \ .$$

In order to count the number of degrees of freedom for $Z(H_{RC})$, we note that both the ε's and the δ's satisfy 2 constraints due to (8.16) and (8.17). Hence there are (I-2) unconstrained ε's and (J-2) unconstrained δ's. The total number of unconstrained parameters is therefore found by adding ρ, the (I-1) τ_i^A's and the (J-1) τ_j^B's to the (I-2) ε's and the (J-2) δ's. The degrees of freedom are accordingly

$$df(H_{RC}) = IJ - 1 - ((I-2) + (J-2) + 1 + (I-1) + (J-1)) = (I-2)(J-2) \ .$$

We thus reject the RC-association model at level α if the observed value $z(H_{RC})$ of $Z(H_{RC})$ satisfies $z(H_{RC}) > \chi^2_{1-\alpha}(df(H_{RC}))$.

For later use note that without restrictions on the 2-factor interactions, ML-estimates for the τ_{ij}^{AB}'s are easily obtained from

$$\ln x_{ij} = \ln \hat{\mu}_{ij} = \hat{\tau}_{ij}^{AB} + \hat{\tau}_i^A + \hat{\tau}_j^B + \hat{\tau}_0$$

using that all τ's sum to zero over all subscripts to get

$$t_{ij}^{AB} = L_{ij} - \bar{L}_{i.} - \bar{L}_{.j} + \bar{L}_{..}, \qquad (8.20)$$

where $L_{ij} = \ln x_{ij}$, a subscript "." means a summation over that subscript and a bar indicates an average.

EXAMPLE 8.2. *Income and wealth.*
From the data base of the Danish Welfare Study, we have selected the personsfor which information on both annual taxable income and wealth is available in the data base and who in addition rented their home, rather than owned it in 1974. The cross-classification of Income (ivided in five income intervals) and Wealth (also divided in five intervals) is shown as Table 8.5.

TABLE 8.5. A random sample of renters in Denmark in 1974 cross-classified according to Income and Wealth.

Income (1000 Dkr.)	Wealth (1000 Dkr.)				
	0	1 - 50	50 - 150	150 - 300	300 -
0 - 40	292	126	22	5	4
40 - 60	216	120	21	7	3
60 - 80	172	133	40	7	7
80 - 110	177	120	54	7	4
110 -	91	87	52	24	25

Source: The data base from the Danish Welfare Study. Cf. Example 3.2.

A test of independence H for the data in Table 8.5 gives a z-value of

$$z(H) = 167.99.$$

With 16 degrees of freedom the independence hypothesis is thus rejected at any reasonable level.

The Z-test statistic for the goodness of fit of an RC-association model has observed value

$$z(H_{RC}) = 14.46$$

with df = 3 · 3 = 9 degrees of freedom. The level of significance is p = 0.107, so we can accept the model.

The ML-estimates for the parameters - except the main effects - of the model are shown in Table 8.6.

TABLE 8.6. ML-estimates for the parameters of the RC-association model - except the main effects.

$\hat{\rho}$:   2.258

	i = 1	2	3	4	5
$\hat{\varepsilon}_i$:	-0.557	-0.339	0.060	0.086	0.751

	j = 1	2	3	4	5
$\hat{\delta}_j$:	-0.636	-0.360	0.062	0.352	0.582

Since ML-estimates for the 2-factor interactions can be derived from Equation (8.20) we can evaluate the fit of the model by comparing these estimates with the products

$$\hat{\rho}\hat{\varepsilon}_i\hat{\delta}_j \qquad (8.21)$$

estimating the right hand side of (8.10). This is done in Table 8.7.

TABLE 8.7. Comparison of the ML-estimates of the 2-factor interactions and the products (8.21).

$\hat{\tau}^{AB}_{ij}$	j=1	2	3	4	5
i=1	0.663	0.246	-0.299	-0.352	-0.258
2	0.431	0.267	-0.276	0.054	-0.476
3	-0.070	0.096	0.095	-0.220	0.098
4	0.025	0.060	0.462	-0.153	-0.395
5	-1.048	-0.669	0.017	0.671	1.030

$\hat{\rho}\hat{\varepsilon}_i\hat{\delta}_j$	j = 1	2	3	4	5
i=1	0.800	0.453	-0.078	-0.443	-0.732
2	0.487	0.276	-0.047	-0.269	-0.446
3	-0.086	-0.048	0.008	0.047	0.078
4	-0.123	-0.070	0.012	0.068	0.113
5	-1.078	-0.610	0.104	0.597	0.987

As can be seen very clearly from the table the estimated products (8.21) match the estimated 2-factor interactions quite well, thus confirming that the RC-association model describes the data well. The sign pattern in Table 8.7, with different signs in the NW/SE corners and in the SW/NE corners, is a consequence of (8.21) and thus a typical feature of the model.

ASSOCIATION MODELS

The parameters ε_i are called **row scores** and the δ_j's are called **column scores**. These names are derived from the fact that they are a scoring of the categories of the row and the columns which, through the product (8.10) defining the RC-association model, describe the dependencies between the row variable and the column variable. Tables 8.6 and 8.7 shows how the estimated scores interact to describe the variation in the 2-factor interactions.

In many studies the normalization of the ε's and the δ's are not (8.12) and (8.13) but instead

$$\sum_i x_{i.}\varepsilon_i = \sum_j x_{.j}\delta_j = 0 .$$

and

$$\sum_i x_{i.}\varepsilon_i^2 = \sum_j x_{.j}\delta_j^2 = 1 .$$

The normalization of the main effects in (8.14) are then also changed to

$$\sum_i x_{i.}\tau_i^A = \sum_j x_{.j}\tau_j^B = 0 .$$

These alternative normalization are preferred by many because they put more weight on categories which are heavily represented in the table. For evaluating the fit of the model it does not matter what normalization we use.

Sometimes there are good reasons to believe that a certain set of known scores for either the row or the column categories will fit the data. If the row scores are known, the model (8.10) for the interactions can then be reformulated as

$$\tau_{ij}^{AB} = \rho e_i \delta_j \qquad (8.22)$$

where the vector $(e_1,...,e_I)$ is a set of known scores. Technically the e's must obey the same normalization (8.12) and (8.13) as the ε's. We may for example believe that the row scores are equidistant. Thus for I=5 both the values (1, 2, 3, 4, 5) and (-2, -1, 0, +1, +2) give the normalized values

(-0.632, -0.316, 0.000, +0.316, +0.632).

The model with known row scores, so that the full model becomes

$$\mu_{ij} = \exp\left(\rho e_i \delta_j + \tau_i^A + \tau_j^B + \tau_0\right) ,$$

is called a **column effects association model**. The column effects model is in contrast to the RC-association model log-linear and ML-estimates can be obtained by the iterative proportional fitting method.

If it is column scores rather than row scores which are known we get the interactions

$$\tau_{ij}^{AB} = \rho \varepsilon_i d_j \qquad (8.23)$$

with $(d_1,...,d_J)$ being the known scores satisfying (8.12) and (8.13) with δ_j replaced by d_j. The full model is then

$$\mu_{ij} = \exp\left(\rho \varepsilon_i d_j + \tau_i^A + \tau_j^B + \tau_0\right) ,$$

It is now called a **row effects association model**. Also this model is log-linear.

Both model (8.22) and model (8.23) can be tested in the usual way by Z-test statistics. The degrees of freedom for the column effects association model are the degrees of freedom for the RC-association model plus (I-2) because (I-2) more parameters are specified, which gives

$$df = (I-2)(J-2) + (I-2) = (I-2)(J-1) .$$

For the row effects association model we get in the same way

$$df = (I-2)(J-2) + (J-2) = (I-1)(J-2) .$$

EXAMPLE 8.2 (continued). *For the data on income and wealth in Table 8.5 it is tempting to try to use the interval midpoints as scores in a row effects model or a column effects model. To do so we have to choose values somewhat arbitrarily for the upper intervals. For income we thus choose the values (20, 50, 70, 95, 130) in units of thousand Danish kroner. After normalization this gives the values*

(-0.630, -0.273, -0.036, +0.261, +0.677).

For wealth we choose (0, 25, 100, 225, 400) also in thousand Danish kroner. Some claim that income and wealth distributions are skew and that the distributions of log-income and log-wealth are more symmetric. If this is the case we should use the interval midpoints on a logarithmic scale. For income this gives the values (1.84, 3.88, 4.23, 4.54, 4.78) and for wealth the values (0, 1.96, 4.46, 5.35, 5.96). For wealth we have to use ln(1)=0 rather than ln(0)=-∞.

In Table 8.8 the Z-test statistic for fit of the RC-association model is compared with the corresponding Z-test statistics for the row effects model and the column effects model for both direct interval midpoint scores and logarithmic midpoint scores.

ASSOCIATION MODELS

TABLE 8.8. The fit of the RC-association model compared with the fit of the row effects model and the column effects model for different midpoint scores.

Model	z	df	Level of significance
RC-association	14.46	9	0.107
Column effects, midpoint scores	23.13	12	0.027
Row effects, midpoint scores	40.28	12	0.000
Column effects, log-midpoint scores	70.67	12	0.000
Row effects, log-midpoint scores	18.44	12	0.103

The conclusion from Table 8.8 is that the column effects model is barely acceptable, which means that known row scores equal to the income interval midpoints is acceptable at a 2% level, but not at a 5% level. A logarithmic transformation of the incomes makes the fit completely unacceptable. The conclusion for wealth is almost the exact opposite. For the direct midpoint scores the fit is totally unacceptable, but if we use interval midpoints on a logarithmic scale, the fit is almost as good as the fit of the RC-association model.

8.5 Correspondence analysis

Correspondence analysis is a statistical technique which has many features in common with the RC-association model.

The RC-association model can be described as a multiplicative model for the 2-factor interactions

$$\tau_{ij}^{AB} = \rho \varepsilon_i \delta_j . \qquad (8.24)$$

Consider now the independence hypothesis H. An alternative way to express H is

$$H : \pi_{ij} - \pi_{i.} \pi_{.j} = 0. \qquad (8.25)$$

If we compare this with the log-linear parameter formulation of independence

$$H : \tau_{ij}^{AB} = 0 ,$$

we can as in (8.24) formulate an alternative to the independence hypothesis by assuming that the left hand side in (8.25) is the product

$$\pi_{ij} - \pi_{i.}\pi_{.j} = \rho \varepsilon_i \delta_j . \tag{8.26}$$

This is essentially the correspondence analysis model, but traditionally (8.26) is rewritten as

$$\left(\frac{\pi_{ij}}{\pi_{i.}\pi_{.j}} - 1 \right) = \rho \varepsilon_i \pi_{i.} \delta_j \pi_{.j} = \lambda \varphi_i \psi_j . \tag{8.27}$$

and in addition more than one product term is allowed on the right hand side in (8.27), giving

$$\left(\frac{\mu_{ij}}{\mu_{i.}\mu_{.j}} - 1 \right) = \sum_{m=1}^{M} \lambda_m \varphi_{im} \psi_{jm} . \tag{8.28}$$

Equation (8.28) defines together with the necessary constraints on the parameters, defined below, the **correspondence analysis model**. Correspondence analysis is, however, seldom thought of as a model, but rather as a data descriptive procedure with scaled parameter estimates as the descriptive tools. For this reason the μ's in (8.28) are replaced by their estimates. Since $\hat{\pi}_{ij} = x_{ij}/n$, (8.28) becomes

$$\left(\frac{x_{ij}/n}{(x_{i.}/n)(x_{.j}/n)} - 1 \right) = \sum_{m=1}^{M} \lambda_m \varphi_{im} \psi_{jm} , \tag{8.29}$$

which is the basis for correspondence analysis.

Another tradition in correspondence analysis is to work with frequencies instead of counts, such that f_{ij} replaces x_{ij}/n in all formulas. This means that (8.29) becomes

$$\left(\frac{f_{ij}}{f_{i.}f_{.j}} - 1 \right) = \sum_{m=1}^{M} \lambda_m \varphi_{im} \psi_{jm} , \tag{8.30}$$

The φ's and the ψ's are subject to the following constraints

$$\sum_i \varphi_{im} f_{i.} = \sum_j \psi_{jm} f_{.j} = 0 , \text{ for all } m , \tag{8.31}$$

$$\sum_i \varphi_{im}^2 f_{i.} = \sum_j \psi_{jm}^2 f_{.j} = 1 , \text{ for all } m \tag{8.32}$$

and

$$\sum_i \varphi_{im}\varphi_{iq}f_{i\cdot} = \sum_j \psi_{jm}\psi_{jq}f_{\cdot j} = 0 \ , \ \text{for all } m \neq q \ . \quad (8.33)$$

Correspondence analysis is very much a matrix based technique. We shall now formulate the basic elements in correspondence analysis in matrix language.

Two key matrices are

$$C_I = \begin{bmatrix} f_{1\cdot} & \cdots & 0 \\ & \cdots & \\ 0 & \cdots & f_{I\cdot} \end{bmatrix}$$

and

$$C_J = \begin{bmatrix} f_{\cdot 1} & \cdots & 0 \\ & \cdots & \\ 0 & \cdots & f_{\cdot J} \end{bmatrix},$$

i.e. diagonal matrices containing the marginal frequencies in the diagonals We further define F as the matrix with elements f_{ij} and the matrix R of residuals as a matrix with elements $r_{ij} = f_{ij} - f_{i\cdot}f_{\cdot j}$. Since (8.30) can be written as

$$f_{ij} - f_{i\cdot}f_{\cdot j} = \sum_{m=1}^{M} f_{i\cdot}\varphi_{im}\lambda_m\psi_{jm}f_{\cdot j} \ , \quad (8.34)$$

correspondence analysis can be defined by the matrix equation

$$R = C_I \Phi \Lambda \Psi' C_J \ ,$$

or

$$C_I^{-1} R C_J^{-1} = \Phi \Lambda \Psi' \ , \quad (8.35)$$

where C_I, C_J, R have already been defined, Φ is a matrix of dimension I×M with elements φ_{im}, Ψ is a matrix of dimension J×M with elements ψ_{jm} and Λ is a diagonal matrix of dimension M with diagonal elements λ_m. Any "solution" to the correspondence analysis problem is thus a question of Equation (8.35) having a set of solutions. This problem is a very old one in mathematics and has the following solution.

For a wide range of matrices S of dimension I×J - and it can be shown that $C_I^{-1}RC_J^{-1}$ is such a matrix - there exists a so-called **single value decomposition** of S, i.e there exist matrices E of dimension I×M, D of dimension J×M and a diagonal matrix Λ of dimension M×M such that

$$S = E \Lambda D', \tag{8.36}$$

where $M = \min\{\text{rank}(E), \text{rank}(D)\}$. The matrices E and D are usually normalized such that

$$E'E = D'D = I_M, \tag{8.37}$$

where I_M is the unit matrix of dimension $M \times M$ with zeros outside the diagonal and 1's in the diagonal. Other normalizations than (8.37) can also be used, however, and in correspondence analysis we have to normalize in a slightly different way.

All this means that we can solve Equation (8.35) by a single value decomposition

$$C_I^{-1} R C_J^{-1} = \Phi \Lambda \Psi'. \tag{8.38}$$

of $C_I^{-1} R C_J^{-1}$.

The normalizations (8.32) and (8.33) can be written in matrix form as

$$\Phi' C_I \Phi = \Psi' C_J \Psi = I_M. \tag{8.39}$$

The normalizations are thus slightly different from those given by (8.37). The single value decomposition in (8.38) gives a complete solution to the correspondence analysis equations (8.30), but there may be many terms on the right hand side of (8.30) and therefore many parameters connected with the row and column categories of the two variables to be compared. If we only want one or two terms we need the following result.

THEOREM 8.1. *Let S be the matrix $C_I^{-1} R C_J^{-1}$ with elements s_{ij} and $Q(M_0)$ the sum of squares*

$$Q(M_0) = \sum_i \sum_j \left(s_{ij} - \sum_{m=1}^{M_0} \lambda_m \varphi_{im} \psi_{jm} \right)^2.$$

Then $Q(M_0)$ is minimized if

$$\lambda_1, \ldots, \lambda_{M_0}$$

are the M_0 largest values in Λ, φ_{im}, $i=1,\ldots,I$ is the column in Φ corresponding to the mth largest λ and ψ_{jm}, $j=1,\ldots,J$ is the column in Ψ corresponding to the mth largest λ, where Φ, Ψ and Λ are the matrices defined by the single-value decomposition (8.38).

With Φ_0 being the $I \times M_0$-dimensional matrix formed by the columns of Φ corresponding to the m largest λ's, Λ_0 being a diagonal matrix with the m largest

λ's in the diagonal and Ψ_0 defined in the same way as Φ_0, theorem 8.1 can be stated as

$$C_I R C_J \approx \Phi_0 \Lambda_0 \Psi_0'.$$

in the sense that the sum of squares between the corresponding elements on the left hand side and the right hand side is minimized.

In a single value decomposition the diagonal elements in the middle matrix are called **eigenvalues** and the columns of the left and right matrix are called **eigen vectors**. It can be shown that all eigenvalues in the decomposition (8.38) are positive. (It is a property of the matrix **R**.) Thus if we, for example, want a correspondence analysis with two terms in (8.29) the parameter estimates for λ_1 and λ_2 are the two largest eigenvalues, and the φ's and the ψ's the corresponding eigenvectors in Φ and Ψ.

As mentioned, correspondence analysis is primarily a data descriptive analysis. There are many ways in correspondence analysis to illustrate the data structure of the contingency table, but the most important is graphical.

A **correspondence analysis diagram** is a diagram in which for $M_0=2$ the elements of the eigenvectors - properly scaled - for dimension two are plotted against the elements of the eigenvectors for dimension one - also properly scaled. Here dimension one is the one corresponding to the largest eigenvalue and dimension two the one corresponding to the second largest eigenvalue. There has been some debate over the scaling. An often used choice for the row categories is to plot the elements of the vector

$$(\lambda_1 \varphi_{11}, \ldots, \lambda_1 \varphi_{I1})$$

as x-coordinates and the elements of the vector

$$(\lambda_2 \varphi_{12}, \ldots, \lambda_2 \varphi_{I2})$$

as y-coordinates. In the same way for the column categories

$$(\lambda_1 \psi_{11}, \ldots, \lambda_1 \psi_{J1})$$

are plotted as x-coordinates and

$$(\lambda_2 \psi_{12}, \ldots, \lambda_2 \psi_{J2})$$

are plotted as y-coordinates.

There has been much debate about how to interpret correspondence analysis diagrams. On this point the reader is advised to consult textbooks on correspondence analysis, cf. the biographical notes. Some of the possible interpretations are discussed in connection with Figure 8.1 below.

EXAMPLE 8.3. *Attending meetings.*
In the Danish Welfare Study one of the questions asked was "How often do you attend meetings outside working hours?". The categories are shown in Table 8.9, where the responses are cross-classified with Social group for 1779 persons with age between 40 and 59.

TABLE 8.9. A random sample of Danes in 1974 with age between 40 and 59 cross-classified according to Frequency of attending meetings outside working hours and Social group.

Social group	Frequency of attending meetings				
	One or more times a week	One or more times a month	Approx. once every second month	A few times a year	Never
I	17	27	13	24	25
II	25	57	17	49	55
III	38	91	41	217	213
IV	22	33	21	133	222
V	9	21	17	87	305

Source: The data base from the Danish Welfare Study. Cf. Hansen (1984).

The correspondence analysis diagram for the data in Table 8.9 is shown as Figure 8.1.

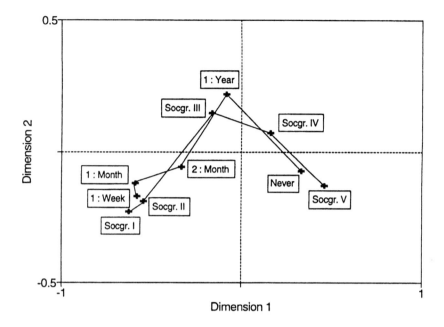

FIGURE 8.1. Correspondence analysis diagram for the data in Table 8.9.

Since a correspondence analysis describes deviations from what should be expected under independence it follows that points on the correspondence analysis diagram which are far from the centre (0.0) indicate categories for which the observed counts deviate most from what should be expected under independence. It can in addition be shown that row categories for which the points in the correspondence analysis diagram are close have the same type of association with the column categories and vice versa. This means that, as regards attending meetings, persons in social groups I and II behave in a similar fashion. In the same way the diagram shows that for all social groups the percentage attending meetings once a week and once a month are about the same. More controversial is the interpretation of closeness of points corresponding to a row category and a column category. Some prominent scholars argue that it is in fact wrong to plot row category points and column category points in the same diagram. It can, however, be argued that if a point for a certain row category is close to a point for a certain column category then the observed numbers in the cell in the contingency table corresponding to these two categories are significantly larger than the expected numbers under independence. The major problem with this type of interpretation is that it critically depends on the scaling of the plotted eigenvectors. If we accept this type of interpretation, however, Figure 8.1 seems to indicate that persons in the two highest social groups attend meetings rather frequently, and that persons in the two lowest social groups almost never attend meetings.

We shall - without going into many more details of correspondence analysis - briefly mention one more central concept. It can be shown that the observed value

$$q = \sum_i \sum_j \frac{\left(x_{ij} - \frac{x_i \cdot x_{\cdot j}}{n}\right)^2}{\frac{x_i \cdot x_{\cdot j}}{n}}$$

of Pearson's test statistic for independence can be rewritten as

$$q = n \sum_i \sum_j \left(\frac{f_{ij}}{f_i \cdot f_{\cdot j}} - 1\right)^2 f_i \cdot f_{\cdot j} \ . \tag{8.40}$$

If we choose to consider only M_0 dimensions, the difference between the left and right hand side of (8.30) for $M = M_0$ would tell us how close we are to the zero we would get if all dimensions M are included. This means that from

$$q_0 = n \sum_i \sum_j \left(\frac{f_{ij}}{f_i \cdot f_{\cdot j}} - 1 - \sum_{m=1}^{M_0} \lambda_m \varphi_{im} \psi_{jm}\right)^2 f_i \cdot f_{\cdot j} \ ,$$

we should get an impression of how much of the dependencies are explained by the correspondence analysis parameters.

Since by Equation (8.30)

$$q_0 = n \sum_i \sum_j \left(\sum_{m=1}^{M} \lambda_m \varphi_{im} \psi_{jm} - \sum_{m=1}^{M_0} \lambda_m \varphi_{im} \psi_{jm}\right)^2 f_i \cdot f_{\cdot j} \ . \tag{8.41}$$

q_0 is close to zero if almost all dependencies are explained. If q_0 is close to q very little of the dependencies have been explained.

From (8.41) follows (using (8.23)) that

$$q_0 = n \sum_{m=M_0+1}^{M} \lambda_m^2$$

or

$$q - q_0 = n \sum_{m=1}^{M_0} \lambda_m^2 \ .$$

We can therefore use

$$r^2(M_0) = \frac{q - q_0}{q}$$

as a measure of how much of the dependencies in the contingency table are accounted for by a correspondence analysis, when a given number M_0 of dimensions included. Almost all the dependencies are accounted for by M_0 dimensions if $r^2(M_0)$ is close to one, and very little has been accounted for if $r^2(M_0)$ is close to zero.

EXAMPLE 8.3 (continued). *For the data in Table 8.9, the eigenvalues and the values of r^2 are shown in Table 8.10.*

TABLE 8.10. Eigenvalues and values of r^2 for the data in Table 8.9.

	Dimension			
	m=1	2	3	4
λ_m	0.354	0.139	0.048	0.021
$r^2(m)$	0.850	0.981	0.997	1.000

Table 8.10 strongly suggests that a model with $M_0 = 2$ fits the data well with almost 98% of the dependencies accounted for, while a model with only one dimension included still has about 15% of the dependencies to account for.

8.6 Bibliographical notes

The symmetry model was first suggested by Bowker (1948) while quasi symmetry was introduced by Caussinus (1965). Caussinus (1965) also noted the connection between marginal homogeneity, symmetry and quasi-symmetry. For 2×2 tables the marginal homogeneity hypothesis and the symmetry hypothesis coincide and lead to the McNemar test suggested by McNemar (1947). The RC-association model was introduced in the form treated here by Goodmann (1979). The model was developed independently by the Danish statistician Georg Rasch, but not published in English, cf. the references in Goodmann (1979). The relationships between the RC-association models and correspondence analysis was pointed out by van der Heijden and de Leeuw (1985), by Goodman (1986) and by van der Heijden, de Falguerolles and de Leeuw (1989).

Goodman's contributions to association models are collected in Goodman (1984). Association models are treated at textbook level in Andersen (1990), chapter 10.

Correspondence analysis has a long history and is known under many names, see Greenacre (1984) or Nishisato (1980). In France it was developed independently by Benzecri (1973). A number of books devoted to correspondence analysis are now available in English, for example Lebart, Morineau and Warwick (1984) and

Greenacre (1984). The geometry of correspondence analysis was discussed by Greenacre and Hastie (1987).

8.5 Exercises

8.1 The two tables below show the forecasts for production and prices for the coming three year periods given by experts in July 1956 and the actual production figures for production and prices in May 1959 given from a sample of 4000 Danish factories.

Prices: Forecast 1956	Actual 1959 Higher	No change	Lower
Higher	209	169	6
No change	190	3073	184
Lower	3	62	81

Production: Forecast 1956	Actual 1959 Higher	No change	Lower
Higher	532	394	69
No change	447	1727	334
Lower	39	230	231

The expected numbers under quasi-symmetry are

Prices: Forecast 1956	Actual 1959 Higher	No change	Lower
Higher	209.00	168.57	6.50
No change	190.50	3073.00	183.50
Lower	2.50	62.50	81.00

Production: Forecast 1956	Actual 1959 Higher	No change	Lower
Higher	532.00	400.33	62.67
No change	440.67	1727.00	340.33
Lower	45.33	223.67	231.00

(a) Test the hypotheses of symmetry and quasi-symmetry on both tables.

(b) Compare the results of the two tests in (a).

ASSOCIATION MODELS

The tables below show the expected numbers under marginal homonegeity.

Prices: Forecast 1956	Higher	Actual 1959 No change	Lower
Higher	209.00	180.07	4.19
No change	178.99	3073.00	123.27
Lower	5.27	122.19	81.00

Production: Forecast 1956	Higher	Actual 1959 No change	Lower
Higher	532.00	413.62	56.52
No change	426.76	1727.00	276.79
Lower	46.39	289.92	231.00

(c) Test if a hypothesis of marginal homogeneity fits none, one, or both of the data sets.

8.2 In connection with the Danish referendum on membership of the European Community (EEC) polls were taken regarding the attitude towards the EEC by the Danish Polling Institute AIM in October 1971 and again in December 1973. The opinions expressed were:

	December 1973		
October 1971	For membership	Against membership	Undecided
For membership	167	36	15
Against membership	19	131	10
Undecided	45	50	20

The expected numbers under quasi-symmetry are

	December 1973		
October 1971	For membership	Against membership	Undecided
For membership	167.00	35.56	15.44
Against membership	19.44	131.00	9.56
Undecided	44.56	50.44	20.00

(a) Test both the hypotheses of symmetry and quasi-symmetry for these data.

Denmark joined the EEC in September 1972 after a referendum with a clear majority for membership.

(b) Interpret the results from (a) given this information.

The expected numbers under marginal homogeneity are given below.

	December 1973		
October 1971	For membership	Against membership	Undecided
For membership	167.00	28.88	27.34
Against membership	25.22	131.00	33.11
Undecided	31.00	29.45	20.00

(c) Test the hypothesis of marginal homogeneity.

(d) If the hypothesis of marginal homogeneity is rejected explain how the expected marginals differ from the observed. Do these differences tell you something important about the Danish populations change in attitude towards the EEC between 1971 and 1973?

8.3 In 1962 and again in 1965 a sample of elderly people were asked to rate their health as being: Good, Neither good nor bad, or Bad. The results of these two self ratings were shown below.

ASSOCIATION MODELS

	Health 1965		
October 1971	Good	Neither	Bad
Good	168	51	9
Neither	42	73	23
Bad	5	17	23

The expected numbers under the hypothesis of marginal homogeneity are shown below.

	Health 1965		
October 1971	Good	Neither	Bad
Good	168.00	46.33	7.14
Neither	46.71	73.00	19.84
Bad	6.76	20.22	23.00

(a) Test the hypothesis of marginal homogeneity.

(b) What can you conclude from the result in (a)?

8.4 One of the first applications of the model later to be known as the RC-association model was to a data set analyzed by Georg Rasch (see bibliographical notes). The data, shown in the table below, are the number of criminal cases dropped by the Police between 1955 and 1958 for male teenagers 15 to 19 years old before the case had led to a verdict.

	Age				
Year	15	16	17	18	19
1955	141	285	320	441	427
1956	144	292	342	441	396
1957	196	380	424	462	427
1958	212	424	399	442	430

The parameters of an RC-association model can be estimated as $\hat{\rho} = 0.328$ and

$\hat{\varepsilon}_i$:	i=1	2	3	4	
	-0.590	-0.372	0.319	0.642	
$\hat{\delta}_j$:	j=1	2	3	4	5
	0.518	0.458	0.042	-0.553	-0.485
$\hat{\tau}_i^A$:	i=1	2	3	4	
	-0.102	-0.092	0.088	0.106	
$\hat{\tau}_j^B$:	j=1	2	3	4	5
	-0.666	0.026	0.107	0.297	0.236

with $\hat{\tau}_0 = 5.801$.

(a) Computed the expected values for each cell and test whether an RC-association model fits the data.

(b) Interpret the estimated parameters of the model.

Suppose we use the years as known scores for the row categories. Normalized according to (8.12) and (8.13) the row scores then become (-0.671, -0.224, 0.224, 0.671). The estimated expected values under this model are shown below

	Age				
Year	15	16	17	18	19
1955	137.48	277.15	329.59	449.55	420.23
1956	150.87	302.07	337.91	425.11	399.04
1957	192.81	383.40	403.42	468.13	441.24
1958	211.84	418.38	414.09	443.21	419.49

(c) Test whether this column effects model fit the data and compare with the result in (a).

8.5 The table below shows Urbanization cross-classified with social rank for the

Association Models

Danish Welfare Study.

	Social group			
Urbanization	I-II	III	IV	V
Copenhagen	45	64	160	74
Cop. Suburbs	99	107	174	90
Three largest cities	57	85	153	103
Other cities	168	287	415	342
Countryside	83	346	361	399

The estimated expected numbers under an RC-association model are

	Social group			
Urbanization	I-II	III	IV	V
Copenhagen	58.41	73.25	132.70	78.64
Cop. Suburbs	93.35	92.37	187.29	96.99
Three largest cities	58.63	90.91	148.90	99.57
Other cities	155.82	292.51	437.51	326.16
Countryside	85.81	339.97	356.60	406.62

(a) Check whether an RC-association model fits the data.

The estimated row and column effects are

$\hat{\varepsilon}_i$:	i=1	2	3	4	5
	-0.573	-0.259	-0.403	0.672	0.564

$\hat{\delta}_j$:	j=1	2	3	4
	-0.539	-0.005	0.445	0.095

(b) Interpret the row and column effects.

(c) Do the parameter estimates in (b) indicate that a row effects association model with Social groups scored with values 1, 2, 3 and 4, properly normalized, is likely to fit the data?

The residuals for the RC-association model (observed minus expected numbers divided by the standard error of the difference) have values:

	Social group			
Urbanization	I-II	III	IV	V
Copenhagen	-3.46	-1.52	3.54	-0.83
Cop. Suburbs	1.49	2.44	-1.76	-1.28
3 largest cities	-0.39	-0.85	0.49	0.54
Other cities	1.98	-0.51	-1.83	1.62
Countryside	-1.08	1.36	0.85	-1.89

(d) Can the RC-association models failure to fit the model be attributed to a few cells? If so describe these cells.

8.6. In one of the publications from the Danish Welfare Study we find the following table showing the connection between income and wealth.

Income (1000 D.kr.)	Wealth (1000 D.kr.)				
	0	0-50	50-150	150-300	Over 300
0-40	360	196	120	79	39
40-60	283	196	134	94	59
60-80	269	197	193	127	61
80-110	286	220	209	122	77
Over 110	193	151	174	154	176

The z-test statistic for goodness of fit of an RC-association model has value z = 19.86.

(a) Determine the degrees of freedom for z and test whether an RC-association model fits the data.

The eigenvalues $(\lambda_1, ... ,\lambda_4)$ turn out to be (0.231, 0.086, 0.021, 0.015).

(b) How much of the dependencies between income and wealth are accounted for by dimensions 1 and 2?

The plot coordinates for a correspondence analysis model with two dimensions included are shown below.

Category	x-coordinate (dimension 1)	y-coordinate (dimension 2)
Row 1	-0.281	0.110
Row 2	-0.117	0.034
Row 3	-0.032	-0.106
Row 4	-0.016	-0.086
Row 5	0.418	0.066
Column 1	-0.208	0.072
Column 2	-0.111	-0.023
Column 3	0.072	-0.132
Column 4	0.189	-0.041
Column 5	0.552	0.135

(c) Draw a correspondence analysis diagram and try to give an interpretation.

8.7 From the data base of the Danish Welfare Study one can extract the following table showing the association between Monthly income and Occupation.

Occupation	Monthly income (D.kr.)				
	0-3192	3193-4800	4801-5900	5901-7488	Over 7489
Independent	108	82	52	44	147
White collar workers	226	242	257	362	423
Blue collar workers	308	359	320	230	75
Pensioners	406	37	9	9	4
Unemployed	61	64	2	2	1
Students	174	9	0	0	0

The eigenvalues $(\lambda_1, ... ,\lambda_4)$ turn out to be (0.571, 0.301, 0.152, 0.063).

(a) How much of the dependencies between income and occupation are accounted for by dimensions 1 and 2?

The plot coordinates for a correspondence analysis model with two dimensions included are shown below.

Category	x-coordinate (dimension 1)	y-coordinate (dimension 2)
Row 1	-0.198	0.357
Row 2	-0.421	0.202
Row 3	-0.107	-0.394
Row 4	1.181	0.150
Row 5	0.505	-0.450
Row 6	1.337	0.210
Column 1	0.810	0.085
Column 2	-0.151	-0.356
Column 3	-0.386	-0.287
Column 4	-0.471	-0.007
Column 5	-0.565	0.555

(b) Draw a correspondence analysis diagram and try to give an interpretation.

Appendix

Solutions and output to selected exercises

Chapter 2

2.1 (a) $\ln f(x|\theta) = x \cdot \ln\theta + \ln(1-\theta)$.

(b) $E[T] = n\theta/(1-\theta) = \Sigma_i x_i \Rightarrow \hat{\theta} = \bar{x}/(1+\bar{x})$

(c) $K(\theta) = -\ln(1-\theta) = -\ln(1-e^\tau)$, $K'(\tau) = e^\tau/(1-e^\tau)$ and $K''(\tau) = e^\tau/(1-e^\tau)^2 \Rightarrow$
$\text{var}[t] = (1-e^\tau)^2/(ne^\tau) = 1/\theta(1-\tau)$.

2.2 (a) $z = -2\ln L(\tau_0) + 2\ln L(\hat{\tau}) = 2\Sigma_i x_i(\hat{t}-\tau_0) + 2\ln(1-\exp\{\hat{t}\}) - 2\ln(1-\exp\{\tau_0\})$.

(b) $\Sigma_i x_i = 17$. $\hat{t} = \ln(1.7/2.7) = -0.463$. $\hat{\theta} = 0.629$.
$\tau_0 = -0.3$. $\theta_0 = 0.741$.

$z = 34(-0.463 - (-0.3)) + 20\ln(1-0.630) - 20\ln(1-0.741) = 1.60$.
$z = 1.65$ df = 1 p = 0.199.

2.5 (a) $\ln L = x\ln\lambda - \ln(x!) - \lambda$.
$K(\lambda) = \lambda = e^\tau$.

(b) $nK'(\lambda) = e^\tau = \Sigma X_i$

$\hat{t} = \ln(\bar{x})$.

(c) $K''(\tau) = e^\tau$.
$\text{var}[\hat{t}] = \exp\{-\tau\} = 1/(n\lambda)$.

(d) $\text{var}[\hat{\lambda}] = \lambda/n$ and $g'(\lambda) = 1/\lambda \Rightarrow \text{var}[\hat{t}] = (1/\lambda^2)\lambda/n = 1/(n\lambda)$.

2.6 (b) $H : \hat{\pi} = 1/10$.

$z(H) = 2\Sigma x[\ln x - \ln(13.7)] = 124.15$ df = 8 [*)] p = 0.000   [*)] One cell has zero count.

(c) $H_1 : \pi_1 = \pi_3 = \pi_4 = \pi_5 = \pi_6 = \pi_{(1)}$

$\pi_7 = \pi_8 = \pi_9 = \pi_{10} = \pi_{(2)}$

(d) $\hat{\pi}_{(1)} = (10+18+19+17+19)/5/137 = 0.121$
$\hat{\pi}_{(2)} = (5 + 0 + 2 + 1)/4/137 = 0.015$
$\hat{\pi}_2 = 46/137 = 0.336$

Decades 1800-1900	1800 -1810	10 -20	20 -30	30 -40	40 -50	50 -60	60 -70	70 -80	80 -90	1890 -1900
Number of executions	10	46	18	19	17	19	5	0	2	1
Expected numbers	16.7	46.0	16.7	16.7	16.7	16.7	2.0	2.0	2.0	2.0

$z(H_1) = 10.93$ df = 6 p = 0.09

2.7 $H : \hat{\pi} = 1/10$.

$z(H) = 2\Sigma x[\ln x - \ln(408.5)] = 25.7$ df = 9 p = 0.002.

2.8 (b) $\tau_1 = \theta_1 + \theta_2$ $\tau_2 = \theta_2$

$H_0 : \tau_1 = \ln(4) - \ln(1) = 2\ln(2)$
  $\tau_2 = \ln(2) - \ln(1) = \ln(2)$

$\theta_1 = \tau_1 - \tau_2 = \ln(2) = \theta_2$.

2.9 (a)

	GG	GB	BB
$n\hat{\pi}$	33	66	33

$z = 6.68$ df = 2 p = 0.035.

(c) $\hat{\theta} = 0.197$.

(d)

	GG	GB	BB
$n\hat{\pi}$	39.5	53.0	39.5

$z = 1.56$ df = 1 p = 0.212.

2.10 $n\pi_{11}(1-\pi_{1.})(1-\pi_{.1}) = n\pi_{11}\pi_{2.}\pi_{.2}$

$n\pi_{22}(1-\pi_{2.})(1-\pi_{.2}) = n\pi_{22}\pi_{1.}\pi_{.1}$.

2.11 $\tau_{11} = \ln(\pi_{11}) - \ln(\pi_{22}) = \ln(\pi_{1.}) + \ln(\pi_{.1}) - \ln(\pi_{2.}) - \ln(\pi_{.2})$
       $= \ln(\pi_{1.}) - \ln(\pi_{2.}) + \ln(\pi_{.1}) - \ln(\pi_{.2}) = \theta_1 + \theta_1$.

$\tau_{12} = \ln(\pi_{12}) - \ln(\pi_{22}) = \ln(\pi_{1.}) + \ln(\pi_{.2}) - \ln(\pi_{2.}) - \ln(\pi_{.2})$
       $= \ln(\pi_{1.}) - \ln(\pi_{2.}) = \theta_1$.

$\tau_{21} = \ln(\pi_{21}) - \ln(\pi_{22}) = \ln(\pi_{2.}) + \ln(\pi_{.1}) - \ln(\pi_{2.}) - \ln(\pi_{.2})$
       $= \ln(\pi_{.1}) - \ln(\pi_{.2}) = \theta_2$.

APPENDIX

Chapter 3

3.1 (b) $\hat{t}_0 = \ln(60) = 4.094$.

$\hat{t}_0^* = \ln(60/480) = -2.079$.

3.2 (a)

$\hat{\tau}^{AB}$	j=1	2
i=1	0.104	-0.104
2	-0.104	0.104

(b)

$\hat{\tau}^B$	j=1	2
	-0.203	0.203

3.3 (a)

$\hat{\tau}^{BC}$	k=1	2
j=1	0.049	-0.049
2	-0.049	0.049

(b)

$\hat{\tau}^C$	k=1	2
	0.220	-0.220

3.4 (a)

τ^{ABC} : 24
τ^{AB} : 12
τ^{AC} : 6
τ^{BC} : 8
τ^{A} : 3
τ^{B} : 4
τ^{C} : 2

———
59

(b) $59 = 60 - 1$

3.5 (a)

$H_2 : B \perp C \mid A$
$H_3 : C \perp A, B$
$H_4^* : C \perp A, B$ & $C = u$
$H_4 : A \perp B \perp C$

(b) $H_2 : \pi_{ijk} = \pi_{ij.}\pi_{i.k}/\pi_{i..}$. $H_3: \pi_{ijk} = \pi_{ij.}\pi_{i.k}$. $H_4^* : \pi_{ijk} = \pi_{ij.}/K$. $H_4 : \pi_{ijk} = \pi_{i..}\pi_{.j.}\pi_{..k}$.

3.6 (a) Formula: $x_{i.k}x_{.jk}/x_{..k}$.

Results:

		C = 1	2
A = 1	B = 1	67.3	15.4
	2	53.7	40.6
2	B = 1	21.7	33.6
	2	17.3	88.4

Model	z(H)
AB, AC, BC	0.12
AC, BC	2.72

3.7 (a)

Model	z(H)	df	Level of sign.
AB, AC, BC	5.38	2	0.068
AC, BC	34.49	4	0.000
AB, BC	9.52	3	0.023
AB, AC	6.29	4	0.178
AC, B	35.23	6	0.000
AB, C	10.27	5	0.068
AB	10.32	6	0.112
A, B, C	39.21	7	0.000
A, B	39.26	8	0.000

(b) Expected numbers under model AB.

Formula: $E[X_{ijk}] = x_{ij.}/K$.

APPENDIX

Results:

A: Response	B: Residence	Sex	
		Male	Female
Yes	Copenhagen	285.0	285.0
	Cities	618.0	618.0
	Countryside	962.5	962.5
No	Copenhagen	62.5	62.5
	Cities	78.0	78.0
	Countryside	108.5	108.5

(c) Formula: $\hat{\tau}^{AB}_{ij} = L_{ij} - \bar{L}_{i.} - \bar{L}_{.j} + \bar{L}_{..}$,

with $L_{ij} = \ln(x_{ij.})$.

Results:

τ^{AB}	Residence		
Non-response	Copenhagen	Cities	Countryside
Yes	-0.203	0.073	0.130
No	0.203	-0.073	-0.130

3.8 (a) and (c)

Model	z(H)	df	Level of sign.
AB, AC, BC	0.19	1	0.660
AC, BC	2.44	2	0.295
AB, AC	1.34	2	0.513
AB, BC	11.36	2	0.003
AC, B	3.13	3	0.372
AB, C	12.05	3	0.007
AC	10.09	4	0.039
A, B, C	13.85	4	0.008
A, C	20.81	5	0.001
C	109.83	6	0.000

Chose AC, B or AC.

(b) AC, B : B⊥A,C ; AC : B⊥A,C & B = u.

3.9 (a)

Model	z(H)	df	Level of sign.
AB, AC, BC	4.67	2	0.097
AB, BC	24.59	4	0.001
AC, BC	28.42	4	0.000
AB, AC	5.63	5	0.131
AB, C	24.62	5	0.372
AC, B	28.44	5	0.000

(c) AB, AC : $B \perp C \mid A$

(d) Expected numbers under model AB, AC:

A: Preference of length	B: Age	C: Sex	
		Male	Female
Less than 2 hours	Under 40	68.4	73.6
	Over 40	77.6	83.4
2½ to 3½ hours	Under 40	60.7	41.3
	Over 40	52.3	35.7
4 hours or more	Under 40	63.8	27.2
	Over 40	25.2	10.8

3.10 (a)

Model	z(H)	df	Level of sign.
AB, AC, BC	5.53	1	0.019
AB, AC	28.14	2	0.000
AB, BC	5.63	2	0.060
AC, BC	89.36	2	0.000
AB, C	28.18	3	0.000
BC, A	89.41	3	0.000

(b) Alcohol abuse is hereditary, if model AB, BC is accepted.

(c) No, if model AB, BC is accepted.

APPENDIX 241

Chapter 4

4.1 (1)

Model	z(h)	df	Level of sign.
ABC, ABD, ACD, BCD	21.33	12	0.046
ABC, ABD, ACD	66.64	24	0.000
ABC, ABD, BCD	48.21	24	0.002
ABC, ACD, BCD	32.84	21	0.005
ABD, ACD, BCD	22.97	16	0.115
ABD, ACD, BC	68.69	28	0.000
ABD, BCD, AC	51.66	28	0.004
ACD, BCD, AB	34.58	19	0.016
ACD, BCD	34.58	20	0.021
ACD, AB, BC, BD	75.98	31	0.000
BCD, AB, AC, AD	60.65	31	0.001
ACD, BC, BD	76.65	32	0.000
BCD, AC, AD	61.33	32	0.001

4.1 (2)

Model	z(h)	df	Level of sign.
ABC, ABD, ACD, BCD	7.35	12	0.834
ABC, ABD, ACD	48.02	24	0.002
ABC, ABD, BCD	26.97	24	0.306
ABC, ACD, BCD	10.85	15	0.763
ABD, ACD, BCD	10.73	16	0.826
ABD, ACD, BC	52.04	28	0.004
ABD, BCD, AC	31.28	28	0.305
ACD, BCD, AB	14.43	19	0.758
ACD, BCD	17.04	20	0.651
ACD, AB, BC, BD	55.61	31	0.004
BCD, AB, AC, AD	34.81	31	0.291
ACD, BC, AC	58.76	32	0.003
BCD, AC, BD	38.02	32	0.214
BCD, AC	175.47	35	0.000
BCD, AD	668.72	36	0.000
AC, AD, BC, BD, CD	79.75	44	0.001

4.2

Model	z(h)	df	Level of sign.
ABC, ABD, ACD, BCD	5.21	6	0.517
ABC, ABD, ACD	15.56	9	0.077
ABC, ABD, BCD	19.23	8	0.014
ABC, ACD, BCD	8.44	12	0.750
ABD, ACD, BCD	15.01	12	0.241
ABC, ACD, BD	19.16	15	0.207
ABC, BCD, AD	22.98	14	0.061
ACD, BCD, AB	18.71	18	0.410
ACD, BCD	135.84	24	0.000
ACD, AB, BC, BD	30.28	21	0.086
BCD, AB, AC, AD	34.17	20	0.025
ACD, AB, BC	80.97	24	0.000
ACD, AB, BD	72.53	24	0.000
ACD, BC, BD	143.97	27	0.000
AB, AC, AD, BC, BD, CD	42.26	23	0.008

APPENDIX

4.3

Model	z(h)	df	Level of sign.
AB, AC, AD, BC, BD, CD	1.71	5	0.888
AB, AC, AD, BC, BD	13.80	6	0.032
AB, AC, AD, BC, CD	8.73	6	0.189
AB, AC, BC, BD, CD	13.14	6	0.041
AB, AC, AD, BD, CD	10.93	6	0.090
AB, AD, BC, BD, CD	2.68	6	0.847
AC, AD, BC, BD, CD	2.20	6	0.900
AC, AD, BC, BD	14.32	7	0.046
AC, AD, BC, CD	9.56	7	0.215
AC, BC, BD, CD	13.96	7	0.052
AC, AD, BD, CD	11.53	7	0.117
AD, BC, BD, CD	3.27	7	0.858
AD, BC, BD	14.85	8	0.062
AD, BC, CD	10.63	8	0.223
BC, BD, CD, A	14.51	8	0.070
AD, BD, CD	12.60	8	0.126
AD, BC	20.86	9	0.013
BC, CD, A	21.87	9	0.009
AD, CD, B	18.59	9	0.029
AD, CD	31.29	10	0.000
AD, B, C	28.79	10	0.001
CD, A, B	29.82	10	0.001

4.4 (a)

Model	z(h)	df	Level of sign.
ABC, ABD, ACD, BCD	0.60	1	0.438
ABC, ABD, ACD	1.23	2	0.540
ABC, ABD, BCD	0.96	2	0.620
ABC, ACD, BCD	2.36	2	0.308
ABD, ACD, BCD	1.18	2	0.555
ABC, ABD, CD	1.55	3	0.670
ABC, BCD, AD	3.09	3	0.378
ABD, BCD, AC	1.57	3	0.667
ABC, ABD	1.70	4	0.791
ABC, AD, BD, DC	6.84	4	0.144
ABD, AC, BC, CD	1.93	4	0.748
ABC, AD, BD	6.93	5	0.226
ABD, AC, BC	2.02	5	0.846
ABD, AC	4.12	6	0.661
ABD, BC	135.20	6	0.000
AB, AC, AD, BC, BD	7.30	6	0.294
ABD, C	136.73	7	0.000
AB, AC, AD, BD	9.35	7	0.229
AB, AC, AD	9.35	8	0.314
AB, AC, BD	10.95	8	0.205
AB, AD, BD, C	141.96	8	0.000
AC, AD, BD	9.39	8	0.311
AB, AC, D	10.95	9	0.279
AB, AD, C	141.96	9	0.000
AC, AD, B	9.39	9	0.402
AC, AD	160.44	10	0.000
AC, B, D	10.99	10	0.358
AD, B, C	142.00	10	0.000
AC, B	28.99	11	0.002
AC, D	162.04	11	0.000
A, B, C, D	143.60	11	0.000

APPENDIX

(c)

A: Stage	B: Operation	C: Survival by 10 years	D: X-ray treatment No	Yes
Early	Radical	No	9.8	16.2
		Yes	40.3	66.3
	Limited	No	1.9	3.1
		Yes	7.7	12.7
Advanced	Radical	No	37.4	61.6
		Yes	7.3	12.0
	Limited	No	7.2	11.8
		Yes	1.4	2.3

(d)

Stage	Survival by 10 years Yes	No
Early	-0.761	0.761
Advanced	0.761	-0.761

4.5 (a)

Model	z(h)	df	Level of sign.
AB, AC, AD, BC, BD, CD	6.66	9	0.672
AB, AC, AD, BC, BD	9.39	10	0.495
AB, AC, AD, BC, CD	37.86	11	0.000
AB, AC, BC, BD, CD	6.75	10	0.749
AB, AC, AD, BD, CD	7.60	11	0.748
AB, AD, BC, BD, CD	10.22	10	0.422
AC, AD, BC, BD, CD	6.89	11	0.808
AC, AD, BC, BD	9.62	12	0.649
AC, AD, BC, CD	38.10	13	0.000
AC, BC, BD, CD	6.99	12	0.858
AC, AD, BD, CD	7.86	13	0.853
AD, BC, BD, CD	10.47	12	0.574
AC, BC, BD	9.67	13	0.721
AC, BC, CD	38.20	14	0.000
AC, BD, CD	7.96	14	0.891
BC, BD, CD, A	10.53	13	0.650
AC, BD	10.20	15	0.870
AC, CD, B	38.73	16	0.001
BD, CD, A	11.50	15	0.716
AC, B, D	40.97	17	0.001
BD, A, C	13.74	16	0.618
BD, A	185.26	17	0.000
BD, C	14.55	17	0.628
A, B, C, D	44.51	18	0.001
BD	186.08	18	0.000
B, C, D	45.32	19	0.001

APPENDIX

(c) In the model BD, A, C.

Age	Been to a movie	
	Yes	No
7- 9	-0.210	0.210
10-12	-0.009	0.009
13-15	0.219	-0.219

4.6

Model	z(h)	df	Level of sign.
AB, AC, AD, BC, BD, CD	16.73	13	0.203
AB, AC, AD, BC, BD	16.95	14	0.259
AB, AC, AD, BC, CD	21.47	14	0.090
AB, AC, BC, BD, CD	20.80	16	0.187
AB, AC, AD, BD, CD	50.34	14	0.000
AB, AD, BC, BD, CD	126.45	16	0.000
AC, AD, BC, BD, CD	113.97	16	0.000
AB, AC, AD, BC	21.80	15	0.113
AB, AC, BC, BD	21.77	17	0.194
AB, AC, AD, BD	50.68	15	0.000
AB, AD, BC, BD	126.87	17	0.000
AC, AD, BC, BD	114.00	17	0.000
AB, AC, BC, D	29.61	18	0.041
AB, AC, BD	54.94	18	0.000
AB, BC, BD	131.14	20	0.000
AC, BC, BD	119.73	20	0.000
AB, AC, BC	34.12	19	0.018
AB, AC, D	62.78	19	0.000
AB, BC, D	138.97	21	0.000
AC, BC, D	127.56	21	0.000

4.7 (a)

Model	z(h)	df	Level of sign.
AB, AC, AD, BC, BD, CD	17.49	16	0.355
AB, AC, AD, BC, BD	146.25	15	0.000
AB, AC, AD, BC, CD	36.08	18	0.007
AB, AC, BC, BD, CD	18.08	18	0.450
AB, AC, AD, BD, CD	20.56	18	0.302
AB, AD, BC, BD, CD	67.77	18	0.000
AC, AD, BC, BD, CD	46.95	17	0.000
AB, AC, BC, BD	151.00	22	0.000
AB, AC, BC, CD	36.40	20	0.014
AB, AC, BD, CD	21.20	20	0.386
AB, BC, BD, CD	72.53	20	0.000
AC, BC, BD, CD	47.26	19	0.000
AB, AC, BD	158.39	24	0.000
AB, AC, CD	43.78	22	0.004
AB, BD, CD	79.74	22	0.000
AC, BD, CD	54.47	21	0.000

(b) Chose AB, AC, BD, CD.

(c) Both graphical and decomposable.

(d) $A \perp D | B,C$ & $B \perp C | A,D$.

APPENDIX

4.8

Model	z(H)	df	Level of sign.	Model	z(H)	df	Level of sign.
ABCD, ACDE	14.19	12	0.289	ACD, BCD, CE	31.29	27	0.259
ABCD, ABCE	13.20	12	0.354	ACD, BCD, AE	130.91	28	0.000
ABCE, ACDE	22.64	12	0.031	ACD, ACE, BC	31.87	27	0.273
ABCD, ABDE	115.59	16	0.000	ACE, AD, BC, BD	55.01	30	0.004
ABCD, BCDE	13.95	12	0.304	ACE, BCD	162.93	27	0.000
ACDE, BCDE	13.47	12	0.336	ACD, ACE, BD	34.72	28	0.178
ABDE, BCDE	29.69	16	0.020	BCD, AD, AE, CE	46.22	30	0.030
ABCE, ABDE	39.01	16	0.001	ACD, BCD	1010.87	30	0.000
ABCE, BCDE	149.77	12	0.000	ACD, BC,CE	38.58	30	0.135
ABDE, ACDE	20.21	16	0.211	AC, AD, BC, BD, CE	61.73	33	0.002
ABCD, ACE	16.44	18	0.562	BCD, AC, CE	169.65	30	0.000
ABCD, BCE	20.14	18	0.325	ACD, BD, CE	41.44	31	0.100
ABCD, ABE	122.45	20	0.000	BCD, AD, CE	46.50	31	0.036
ABCE, ACD	24.89	18	0.128	ACD, BC	1018.16	33	0.000
ABCE, ABD	45.86	20	0.001	AC, AD, BC, CE	66.27	34	0.001
ABCE, BCD	155.96	18	0.000	AC, BC, CD, CE	176.94	33	0.000
ACD, ACE, BCD, BCE	21.51	21	0.428	ACD, CE	139.38	33	0.000
ABD, ABE, ACD, ACE	28.89	24	0.224	AD, BC, CD, CE	53.79	34	0.017
ABD, ABE, BCD, BCE	37.57	24	0.038				
ABCD, CE	23.16	21	0.336				
ABCD, AE	122.78	22	0.000				
ABC, ACD, ACE	28.14	24	0.254				
ABC, ABD, ACE	49.10	26	0.004				
ABC, ACE, BCD	159.20	24	0.000				
ACD, ACE, BCD	24.58	24	0.429				
ABD, ACD, ACE	29.26	26	0.300				
ABD, BCD, AE, CE	40.73	28	0.057				

Chapter 5

5.1 (a) AB, AC, BC: 8 - 8 + 0 = 0.

(b) AB, AC: 8 - 6 + 0 = 2. AB, BC: 8 - 6 + 0 = 2.

5.2 (a) AB, AC, BC: 21 - 19 + 2 = 4.

(b) AB, AC: 21 - 15 + 2 = 8. AB, BC: 21 - 15 + 0 = 6. AC, BC: 21 - 15 + 2 = 8.

5.3 (a) AB, AC, BC: 18 - 18 + 2 = 2.

(b) AB, AC: 18 - 16 + 2 = 4. AB, BC: 18 - 12 + 2 = 8. AC, BC: 18 - 15 + 0 = 3.

Chapter 6

6.1

Model	z(H)	df	Level of sign.
AB, AC, BC	5.38	2	0.068
AB, BC	9.52	3	0.023
AC, BC	34.49	4	0.000
A, BC	38.46	5	0.000

6.2

Model	z(H)	df	Level of sign.
AB, AC, BC	0.19	1	0.660
AB, BC	1.34	2	0.513
AC, BC	11.36	2	0.003
A, BC	12.05	3	0.007

6.3

Model	z(H)	df	Level of sign.
BC, AC, AB	5.53	1	0.019
BC, AB	1.63	2	0.060
AC, AB	28.14	2	0.000
C, AB	28.18	3	0.000

APPENDIX

6.4 (1)

Model	z(H)	df	Level of sign.
AB, AC, AD, BCD	34.81	31	0.291
AB, AC, BCD	175.16	34	0.000
AB, AD, BCD	668.69	35	0.000
AC, AD, BCD	38.02	32	0.214
AB, BCD	761.97	38	0.000
AC, BCD	175.47	35	0.000
AD, BCD	668.72	36	0.000
A, BCD	766.74	39	0.000

6.4 (2)

Model	z(H)	df	Level of sign.
AB, AC, AD, BCD	60.65	31	0.001
AB, AC, BCD	78.99	34	0.000
AB, AD, BCD	159.92	35	0.001
AC, AD, BCD	61.30	32	0.000
AB, BCD	185.33	38	0.000
AC, BCD	79.01	35	0.000
AD, BCD	162.42	36	0.000
A, BCD	185.99	39	0.000

6.5

Model	z(H)	df	Level of sign.
DA, DB, DC, ABC	1.37	4	0.849
DA, DB, ABC	13.52	5	0.019
DA, DC, ABC	8.45	5	0.133
DB, DC, ABC	12.86	5	0.025
DA, ABC	19.18	6	0.004
DB, ABC	24.43	6	0.000
DC, ABC	20.22	6	0.003
D, ABC	30.41	7	0.000

6.6

Model	z(H)	df	Level of sign.
CA, CB, CD, ABD	1.93	4	0.748
CA, CB, ABD	2.02	5	0.846
CA, CD, ABD	4.03	5	0.545
CB, CD, ABD	134.13	5	0.000
CA, ABD	4.12	6	0.661
CB, ABD	135.20	6	0.000
CD, ABD	135.67	6	0.000
C, ABD	136.73	7	0.000

6.7

Model	z(H)	df	Level of sign.
DA, DB, DC, ABC	5.29	7	0.624
DA, DB, ABC	8.02	8	0.432
DA, DC, ABC	36.49	9	0.000
DB, DC, ABC	5.37	8	0.717
DA, ABC	38.77	10	0.000
DB, ABC	8.06	9	0.529
DC, ABC	36.59	10	0.000
D, ABC	38.83	11	0.000

APPENDIX

6.8

Model	z(H)	df	Level of sign.
AB, AC, AD, BCD	15.80	12	0.201
AB, AC, BCD	16.44	14	0.287
AB, AD, BCD	66.12	14	0.000
AC, AD, BCD	45.30	13	0.000
AB, BCD	70.88	16	0.000
AC, BCD	45.61	15	0.000
AD, BCD	99.72	15	0.000
A, BCD	106.99	17	0.000

6.9

Model	z(H)	df	Level of sign.
EA, EB, EC, ED, ABCD	23.00	18	0.191
EA, EB, EC, ABCD	23.08	19	0.234
EA, EB, ED, ABCD	122.22	20	0.000
EA, EC, ED, ABCD	23.00	19	0.238
EB, EC, ED, ABCD	23.04	19	0.236
EA, EB, ABCD	122.60	21	0.000
EA, EC, ABCD	23.08	20	0.285
EA, ED, ABCD	122.42	21	0.000
EB, EC, ABCD	23.16	20	0.281
EB, ED, ABCD	122.45	21	0.000
EC, ED, ABCD	23.04	20	0.287
EA, ABCD	122.78	22	0.000
EB, ABCD	122.73	22	0.000
EC, ABCD	23.16	21	0.336
ED, ABCD	122.64	22	0.000
E, ABCD	122.90	23	0.000

Chapter 7

7.1   Set 1: No solutions.
      Set 2: Solutions.
      Set 3: Solutions.

7.2 Set 1: No solutions.
     Set 2: Solutions.
     Set 3: No solutions.

7.3 (a)

Variable	Parameter		Standard error
Intercept	β_0	-2.578	0.087
Sex	β_1	-0.322	0.078
Empl. sector	β_2	0.305	0.074
Residence	β_3	-0.039	0.084

(b)

Variables included	z(H)
All	0.65
Sex, Empl. sector	0.87
Sex	17.14
None	28.80

7.4 (a) $z = 0.49$ df = 1 p = for logistic regression model.

(b)

Variable	Parameter	Estimate	Standard error
Intercept	β_0	2.286	1.514
Dust	β_1	-8.984	1.292
Ventilation	β_2	-0.064	0.169

Variables included	z(H)
Both	0.49
Dust	0.63
Ventilation	51.49
None	81.02

Dust is included, Ventilation excluded.

(c) If City Hall no. 120 is omitted the explanatory variables are almost quasi complete separated.

7.6 (a) Variables B, C and D.

(b)

Variables included	z
B C D E F	0.00
B C D E	0.28
B C D	1.41
B C	6.61
C	16.65

(c) P = 0.98, P = 0.72.

7.7 (a) z = 19.36 df = 16 for a logistic regression model.

(b)

Variable	Parameter		Standard error
Intercept	β_0	0.613	0.431
Sex	β_1	0.198	0.250
Age	β_2	-0.050	0.011

Age is significant. Sex is insignificant.

(c) Dummy variable Age1 to Age9: z(H) = 7.24 df = 8.

Variables included	z
Sex	41.81
Age1 to Age9	7.87
None	42.13

Variable	Parameter		Standard error
Intercept	β_0	-1.481	0.242
Sex	β_1	0.205	0.259
Age1	β_2	-0.460	0.982
Age2	β_3	1.483	0.306
Age3	β_4	0.832	0.306
Age4	β_5	0.197	0.333
Age5	β_6	-0.366	0.368
Age6	β_7	-0.167	0.412
Age7	β_8	0.139	0.400
Age8	β_9	-0.501	0.465
Age9	β_{10}	-0.509	0.513

7.8 (a)

Model	df
A B C D	87
B C D	1
A C D	1
A B D	1
A B C	1
C D	2
B D	2
B C	2
C	3
B	3
None	4

(b) B and C are significant.

APPENDIX 257

(c)

Explanatory variables included	Explanatory variables excluded	df
A B1 B2 B3 B4 C D	None	84
B1 B2 B3 B4 C D	A	1
A B1 B2 B3 B4 D	C	1
A B1 B2 B3 B4 C	D	1
A C D	B1 B2 B3 B4	4
B1 B2 B3 B4 D	A C	2
B1 B2 B3 B4 C	A D	2
C D	A B1 B2 B3 B4	5
B1 B2 B3 B4	A C D	3
C	A B1 B2 B3 B4 D	6

Chapter 8

8.1 (a) Prices:

Symmetry: $z(H) = 62.73$ df = 3 p = 0.000.

Quasi-symmetry: $z(H) = 0.14$ df = 1 p = 0.707.

Production:

Symmetry: $z(H) = 31.07$ df = 3 p = 0.000.

Quasi-symmetry: $z(H) = 2.03$ df = 1 p = 0.154.

(c) Prices: $z(H) = 65.41$ df = 2 p = 0.000.

Production: $z(H) = 29.00$ df = 2 p = 0.000.

8.2 (a) Symmetry: $z(H) = 50.15$ df = 3 p = 0.000.

Quasi-symmetry: $z(H) = 0.06$ df = 1 p = 0.812.

(c) $z(H) = 49.62$ df = 2 p = 0.000.

(d) A number of undecided have changed to No. The nunmber of Yes is almost the same.

8.3 (a) $z(H) = 2.92$ df = 2 p = 0.232.

258 APPENDIX

8.4 (a)

| | Age | | | | |
Year	15	16	17	18	19
1955	139.19	281.22	330.61	446.94	416.36
1956	145.92	293.56	335.02	434.68	406.33
1957	196.42	389.86	404.89	461.24	435.86
1958	211.21	416.57	413.97	443.73	421.44

$z(H) = 3.49$ df = 6 $p = 0.746$.

(c) $z(H) = 4.74$ df = 8 $p = 0.785$.

8.5 (a) $z(H) = 18.16$ df = 6 $p = 0.006$.

8.6 (a) df = 9 $p = 0.019$.

(b) Dimension 1 : 86.9%
    Dimension 1+2: 98.9%

(c)

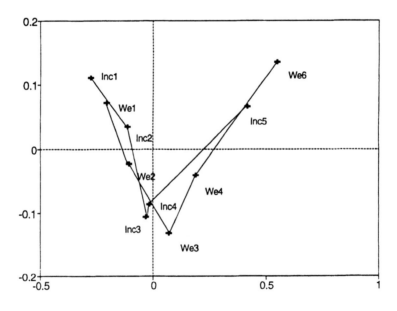

APPENDIX

8.7 (a) Dimension 1 : 73.5%
   Dimension 1+2: 93.9%

(b)
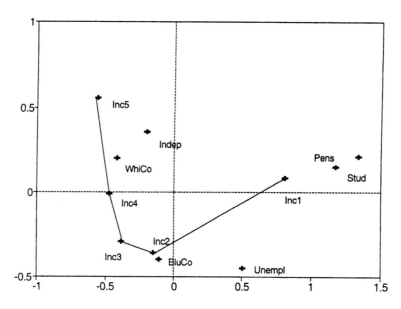

References

Agresti, A. (1990). *Categorical Data Analysis*. New York: J. Wiley and Sons.

Agresti, A. (1996). *An Introduction to Categorical Data Analysis*. New York: J. Wiley and Sons.

Albert, A. and Andersson, J.A. (1984). On the existence of maximum likelihood estimates in logistic regressions models. *Biometrika.* **71**, 1-10.

Andersen, A. H. (1974). Multidimensional contingency tables. *Scand.Jour.Statist.* **3**, 115-127.

Andersen, E.B. (1990). *The Statistical Analysis of Categorical Data*. Heidelberg: Springer Verlag.

Andersen, E.B., Jensen, N.E. and Kousgård, N. (1987). *Theoretical Statistics for Economics, Business Adminstration and the Social Sciences*. Heidelberg: Springer Verlag.

Barndorff-Nielsen, O. (1978). *Information and Exponential Families in Statistical Theory*. New York: J. Wiley and Sons.

Bartholomew, D.J. (1980). Factor analysis for categorical data. *Jour.Royal Statist.Soc.B.* **42**, 293-321.

Benzecri, J.-P. (1973). *L'Analyse des Données. Tome 2: L'Analyse des Correspondances*. Paris: Dunod.

Berkson, J. (1944). Application of logistic functions to bio-assay. *Jour.Amer. Statist.Assoc.* **39**, 357-365.

Berkson, J. (1953). A statistically precise and relatively simple method of estimating the bio-assay with quantal response, based on the logistic function. *Jour.Amer.Statist.Assoc.* **39, 357-365**.

Birch, M.W. (1963). Maximum likelihood in three-way contingency tables. *Jour. Royal Statist.Soc., B.* **25**, 220-233.

Bishop, Y.M.M., Feinberg, S.E. and Holland, P.W. (1975). *Discrete Multivariate Analysis. Theory and Practice*. Cambridge: MIT-Press.

Bowker, A.H. (1948). A test for symmetry in contingency tables. *Jour.Amer. Statist.Assoc.* **43**, 572-574.

Caussinus, H. (1965): Contribution a l'analyse statistique des tableaux de correlation. *Ann.Fac.Sci.Univ. Toulouse.* **29**, 77-182.

Christensen, R. (1990). *Log-Linear Models.* New York: Springer Verlag.

Cox, D.R. (1970). *The Analysis of Binary Data.* London: Methuen and Co.

Darroch, J.N., Lauritzen, S.L. and Speed, T.P. (1980). Markov fields and log-linear interaction models for contingency tables. *Annals Statist.* **8**, 522-539.

Deming, W.E. and Stephan, F.F. (1940). On the least squares adjustment of a sampled frequency table when the expected marginal totals are known. *Annals Math.Statist.* **11**, 427-444.

Edwards, D. (1995). *Introduction to Graphical Modeling.* Heidelberg: Springer Verlag.

Edwards, D. and Kreiner, S. (1983). The analysis of contingency tables by graphical models. *Biometrika.* **70**, 553-565.

Edwards, D. and Havranek, T. (1985). A fast procedure for model search in multidimensional contingency tables. *Biometrika*, **72**, 339-351.

Edwards, D. and Havranek, T. (1987). A fast model selection procedure for large families of models. *Jour.Amer.Statist.Assoc.* **82**, 205-213.

Fienberg, S.E. and Mason, W.M. (1979): Identification and estimation of age-period-cohort models in the analysis of discrete archival data. *Sociol. Methodology.* 1-67.

Goodman, L.A. (1968). The analysis of cross-classified data: Independence, quasi-independence and interactions in contingency tables with and without missing entries. *Jour.Amer.Statist.Assoc.* **63**, 1091-1131.

Goodman, L.A. (1970). The multiplicative analysis of qualitative data: Interactions among multiple classifications. *Jour.Amer.Statist.Assoc.,* **65**, 226-256.

Goddman, L.A. (1971). Some multiplicative models for the analysis of cross-classified data. *Sixth Berkeley Symposium on Probability and Mathematical Statistics.* **I**, 649-696.

Goodman, L.A. (1972). A general model for the analysis of surveys from panel studies and other kinds of surveys. *Amer.Jour.Sociol.* **77**, 57-109.

Goodman, L.A. (1973). Causal analysis of data. *Amer.Jour.Sociol.* **78**, 173-229.

Goodman, L.A. (1978). *Analyzing Qualitative/Categorical Data. Log-linear Models and Latent Structure Analysis.* London: Addisson and Wesley.

Goodman, L.A. (1979). Simple methods for the analysis of association in cross-classifications having ordered categories. *Jour.Amer.Statist.Assoc.* **76**, 320-334.

Goodman, L.A. (1981). Association models and canonical correlation in the analysis of cross-classifications having ordered categories. *Jour.Amer. Statist.Assoc.* **76**, 320-334.

Goodman, L.A. (1984). *The Analysis of Cross-classified Data Having Ordered Categories.* Cambridge: Harvard University Press.

Goodman, L.A: (1986). Some useful extensions of the usual correspondence analysis approach and the usual log-linear models approach in the analysis of contingency tables. *Int.Statist.Review.* **54**, 243-309.

Greenacre, M.J. (1984). *Theory and Applications of Correspondence Analysis.* New York: Academic Press.

Haberman, S.J. (1974). *The Analysis of Frequency Data. Vol. I.* New York: Academic Press.

Haberman, S.J. (1974). *The Analysis of Frequency Data. Vol. II.* New York: Academic Press.

Hansen, E.J. (1978). *The Distribution of Living Conditions. Main Results from the Welfare Survey.* (In Danish). Danish National Institute of Social Research. Publication 82. Copenhagen: Teknisk Forlag.

Hansen, E.J. (1984). *Social Groups in Denmark.* (In Danish). Study No.48. Danish National Institute of Social Research. Copenhagen: Teknisk Forlag.

Holm, S. (1979). A simple sequentially rejective multiple test procedure. *Scand. Jour.Statist.* **6**, 65-70.

Hosmer, D.W. and Lemeshow, S. (1989). *Applied Logistic Regression.* New York: J. Wiley and Sons.

Lauritzen, S.L. (1996). *Graphical Models.* Oxford: Oxford University Press.

Lebart, L., Morineau, A. and Warwick, K. (1984). *Multivariate Descriptive Statistical Analysis.* New York: J. Wiley and Sons.

Lehmann, E.L. (1959). *Testing Statistical Hypotheses.* New York: J. Wiley and Sons.

McCullagh, P. (1980). Regression models for ordinal data. *Jour.Royal Statist.Soc. B.* **42**, 109-142.

McCullagh, P. and Nelder, J.A. (1983). *Generalized Linear Models.* London: Chapman and Hall.

McNemar, H. (1947). Note on the sampling error of the differences between correlated proportions or percentages. *Psychometrika.* **12**, 153-157.

Nelder, J.A. and Wedderburn, R.W.M. (1972). Generalized linear models. *Jour. Royal Statist.Soc. A.* **135**, 370-384.

Nishisato, S. (1980). *Analysis of Categorical Data: Dual Scaling and Its Applications.* Toronto: University of Toronto Press.

Pregibon, D. (1981). Logistic regression diagnostics. *Annals Statist.* **9**, 705-724.

Rao, C.R. (1973). *Linear Statistical Inference and its Applications, 2.ed.* New York: J. Wiley and Sons.

Schaffer, J.P. (1986). Modified sequentially rejective multiple test procedures. *Jour.Amer.Statist.Assoc.* **81**, 826-831.

Thompson, W.A. (1977). On the treatment of grouped observations in life studies. *Biometrics.* **33**, 436-470.

van der Heijden, P.G.M. and de Leeuw, J. (1985). Correspondence analysis: A complement to log-linear analysis. *Psychometrika.* **50**, 429-447.

van der Heijden, P.G.M., de Falguerolles, A. and de Leeuw, J. (1989). A combined approach to contingency table analysis using correspondence analysis and log-linear analysis. *Appl.Statist.* **38**, 249-292.

Weisberg, S. (1985). *Applied Linear Regression. 2 ed.* New York: J.Wiley and Sons.

Whittager, J. (1990). *Graphical Models in Applied Multivariate Statistics.* New York: J. Wiley and Sons.

Subject Index

Association diagram
  for 4-way tables 92
Association model 204
Association diagram 48

Base line category 185
Binomial distribution 12
Bonferroni-procedure 68

Canonical parameter 7
  for binomial distribution 12
  for multinomial distribution 14
  for Poisson distribution 20
Case in logistic regression 158
Categorical variables 42
Clique 94
Column effects association model 215
Column scores 215
Composite hypothesis 21
Conditional logistic regression model 188
Constrained canonical parameters
  hypotheses for 22
Continuation regression model 188
Cook's distance 106, 177
Correspondence analysis diagram 221
Correspondence analysis model 218

Decomposable model 95
Deming-Stephan method 35
Design matrix 29
Deviance 177
Diagnostics 74, 177
Dimension of exponential family 7
Dimension of a contingency table 84
Domain of exponential family 10
  for binomial distribution 13
Dummy variable method 184

Eigenvalue 221
Eigenvector 221
Exact level of significance 134

Explanatory variable 141, 158
Exponential family 6,7

Generalized linear model 194
Geometric distribution 37
Graphic model 94
Grouping of cells 133

Hat matrix 31, 105, 177
Hierarchical model 89
Hierarchically ordered models 91
Hierarchically ordered hypotheses 71

Incomplete tables 54, 127
Interactions 45, 85
Iterative proportional fitting method 35, 53
Iterative weighted least square method 194

Joint point probability 6

Leverage 106, 178
Likelihood equations 8
  for three-way tables 50
  for multiway-tables 88
  for symmetry tables 205
  for RC-association models 210
  for logistic regression 159
Likelihood ratio test 21, 23
Log-likelihood function 8, 14, 21
Log-linear model 15, 29
  for three-way table 42
  for multi-way table 84
Log-linear parametrization 44
Logistic regression 157
Logistic regression model 157
Logistic transformation 157
Logit function 142, 155
Logit model 151

Main effect 45, 85
Marginal homogeneity 209
Maximum likelihood estimates 8

Subject Index

McNemars test statistic 210
Monte Carlo technique 134
Multinomial distribution 4, 6, 13
Multinomial model 84

Newton-Raphson method 162

Parameters 6
Parametric multinomial distribution 15
Partial logistic regression model 188
Point probability 6
Poisson distribution 6, 20
Poisson model 84
Polytomous response variable 187
Prediction
   in logistic regression 182
Prediction probability 142
   for the logit model 147

Quasi-complete separation 161
Quasi-symmetry 205

Random zero 127
RC-association model 210
Reparametrization 15, 22
Reponse variable 141, 158
Residual 4, 73
Residual diagram
   in logistic regression 167
Residual diagram 163
Response probability 158
Response variable 158
Row effects association model 215
Row score 215

Saturated model 29, 46, 87
Sequential Bonferroni-procedure 68
Sequential testing 99, 162, 173
Single value decomposition 219
Square contingency table 204
Standardized estimates 75, 146
Standardized interactions 105
Standardized residual 4, 73
   in logistic regression 167
Statistical model 3

Structural zero 127
Sufficient marginal 52
Sufficient statistic 7
   for binomial distribution 12
   for multinomial distribution 13
   for Poisson distribution 20
Support 10
   for binomial distribution 13
   for Poisson distribution 20
   in logistic regression 160
Survey 85
Symmetry model 204

Three-factor interaction 45
Three-way contingency table 42
Two-factor interaction 45
Two-way contingency table 1
Two-way table 1

Druck: STRAUSS OFFSETDRUCK, MÖRLENBACH
Verarbeitung: SCHÄFFER, GRÜNSTADT